W9-BXS-766

This Is Climate Change

Originally published in the UK as *Small Gases, Big Effect* by Penguin Books Ltd., London, in 2021.
First published in North America in revised form by The Experiment, LLC, in 2021.

The Experiment, LLC
220 East 23rd Street, Suite 600
New York, NY 10001-4658
theexperimentpublishing.com

Library of Congress Cataloging-in-Publication Data

Names: Nelles, David, author. | Serrer, Christian, author.
Title: This is climate change : a visual guide to the facts — see for yourself how the planet is warming and what it means for us / David Nelles and Christian Serrer ; illustrations and infographics by Lisa Schwegler, Stefan Kraiss, Janna Geisse.
Other titles: Small Gases, Big Effect. English
Description: New York : The Experiment, 2021. | Originally published in the UK as Small Gases, Big Effect by Penguin Books Ltd., London, in 2021.
Identifiers: LCCN 2021013663 (print) | LCCN 2021013664 (ebook) | ISBN 9781615198269 | ISBN 9781615197552 (ebook)
Subjects: LCSH: Climatic changes.
Classification: LCC QC903 .N4513 2021 (print) | LCC QC903 (ebook) | DDC 363.738/74--dc23
LC record available at https://lccn.loc.gov/2021013663
LC ebook record available at https://lccn.loc.gov/2021013664

ISBN 978-1-61519-826-9
Ebook ISBN 978-1-61519-755-2

Cover and text design by Jack Dunnington
Author photograph by Edmund Möhrle Photography

Manufactured in Turkey

First printing August 2021
10 9 8 7 6 5 4 3 2 1

This Is Climate Change

A Visual Guide to the Facts

See for Yourself How the Planet Is Warming and What It Means for Us

David Nelles and Christian Serrer

Illustrations and infographics by
Lisa Schwegler, Stefan Kraiss, and Janna Geisse

NEW YORK

Contents

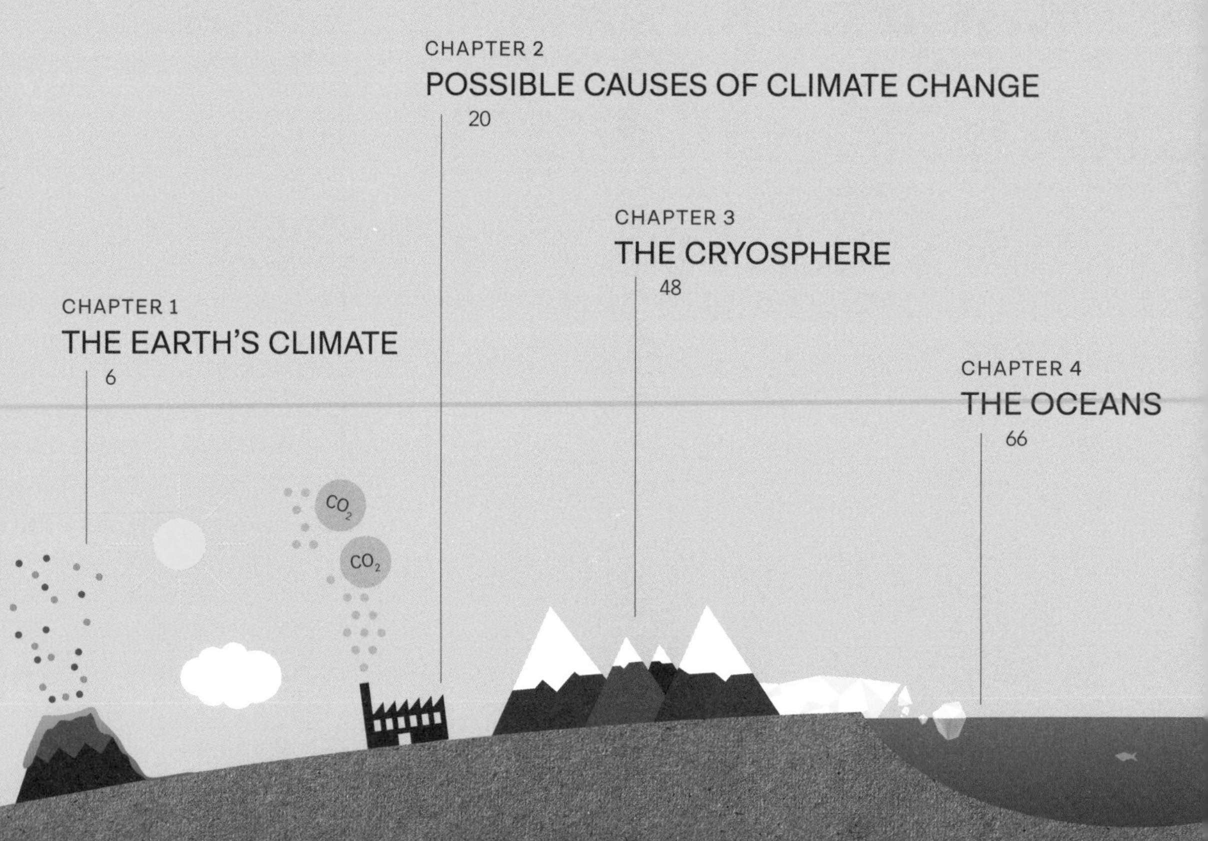

HOW TO READ THIS BOOK

Superscript numbers at the end of sentences (for example,[5]*) are source references. We explain on page 126 how to find the references cited.*

Numbers in a circle (for example, ❶) link the text with the graphics on the page, and appear within the text wherever is appropriate.

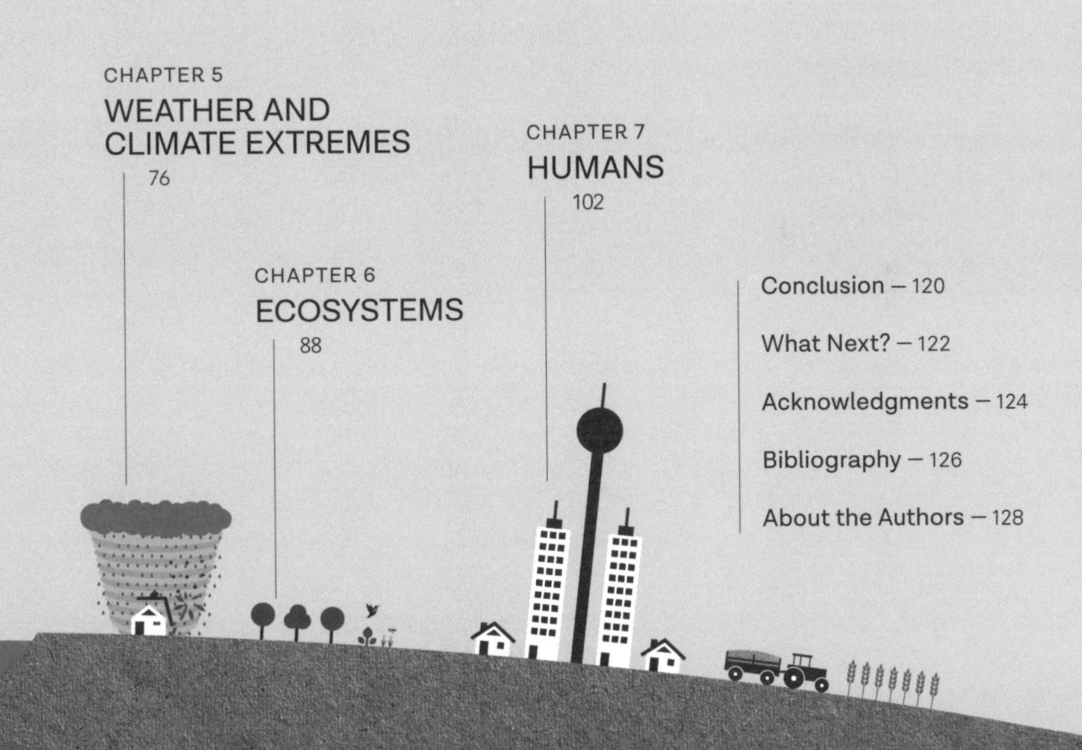

CHAPTER 1

THE EARTH'S CLIMATE

"Climate" refers to the average weather conditions over a long period of time—a period of at least thirty years, according to the World Meteorological Organization.[1] In contrast to the weather, therefore, the climate changes very slowly. A 5°C (9°F) drop in temperature from one day to the next is completely different from the climate cooling by 5°C (9°F). In the latter case, we would be plunged into conditions resembling the last ice age, and North America and Northern Europe would once again be covered by thick sheets of ice.[2]

The Natural Greenhouse Effect

Most of the sun's rays penetrate Earth's atmosphere and reach its surface ❶. These rays are absorbed by the ground and then released as thermal radiation, or heat ❷.[1] Without the presence of certain gases in Earth's atmosphere, such as water vapor (H_2O), carbon dioxide (CO_2), ozone (O_3), nitrous oxide (N_2O) and methane (CH_4), this thermal radiation would simply escape unhindered back into space ❸.[2] This would make the climate around 33°C (59.4°F) colder, and the entire planet would freeze over.[3,4] But, thanks to this layer of naturally occurring gases, thermal radiation is prevented from freely escaping Earth's atmosphere.[5] Instead, a portion is absorbed and then released again, in all directions—including back down toward the surface of Earth ❹.[4] This means that both the lower-lying layers of air and the Earth's surface get heated again.[6] This process of natural warming is known as the "natural greenhouse effect."[2] The gases responsible for this are called "natural greenhouse gases," and they ensure that the average global temperature remains at around 14°C (57°F).[7]

③
①
②
–19°C (–2°F)
CH4
O3
N2O
CO2
H2O
N2O
CO2
O3
H2O
CH4
O3
CH4
N2O
N2O
①
④
14°C (57°F)
②

Natural Greenhouse Gases

In Earth's atmosphere, dry air (i.e., air free of water vapor) consists mainly of nitrogen and oxygen ❶.[1] The concentration of natural greenhouse gases is actually vanishingly small. Carbon dioxide (CO_2), ozone (O_3), nitrous oxide (N_2O) and methane (CH_4) together make up only about 0.04 percent of Earth's atmosphere.[2] Water vapor (H_2O) makes up on average around 0.25 percent.[3]

Despite their low concentrations, however, natural greenhouse gases have a decisive influence on the climate. Unlike oxygen and nitrogen, they can absorb thermal radiation, and thereby prevent it from escaping directly from Earth into space (p. 8).[4] Without them, the climate would be 33°C (59.4°F) colder ❷ and Earth would be uninhabitable.[5,6]

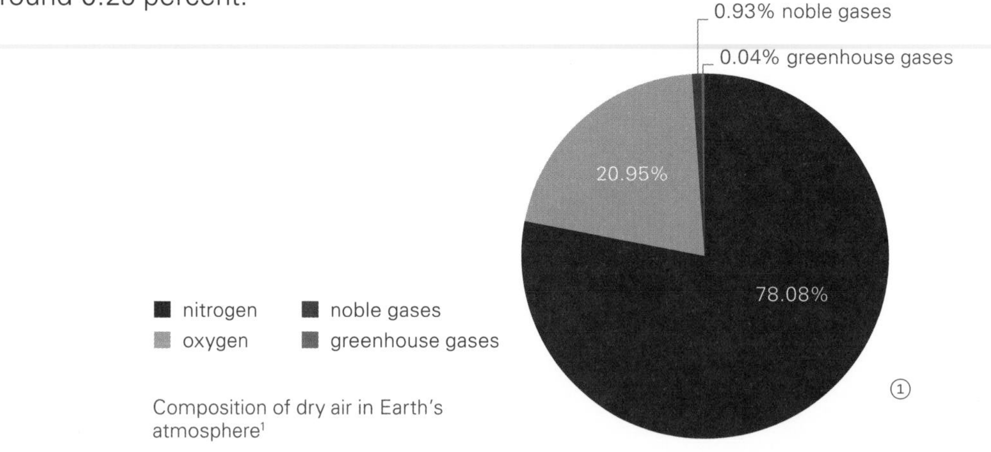

Composition of dry air in Earth's atmosphere[1]

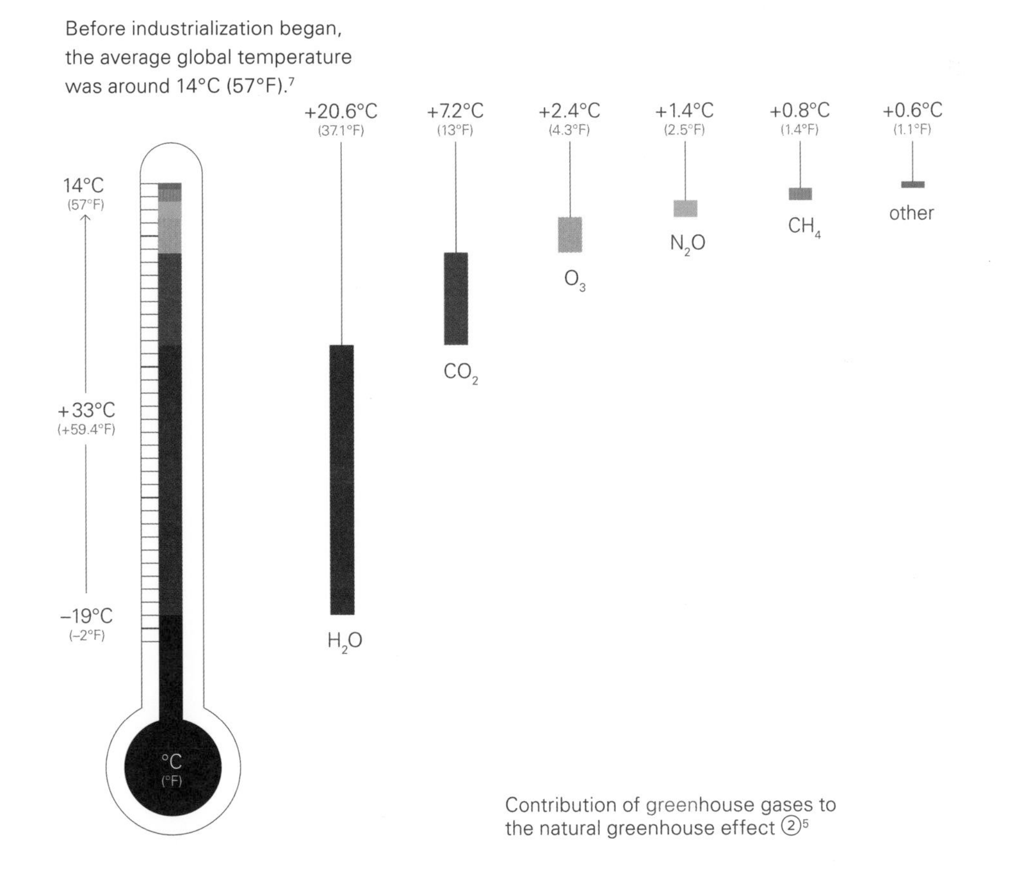

Contribution of greenhouse gases to the natural greenhouse effect ②[5]

Volcanoes and the Sun

Volcanic eruptions can both warm and cool Earth's climate. Volcanic activity releases carbon dioxide (CO_2), which intensifies the greenhouse effect (p. 8), influencing the climate for centuries or even millions of years ❶.[1,2] Eruptions can also spew gases and particles into the upper atmosphere, where they form "aerosols" (p. 26). These aerosols scatter some of the solar radiation back into space ❷,[3] which can cool the climate for several years after the eruption.[2,4]

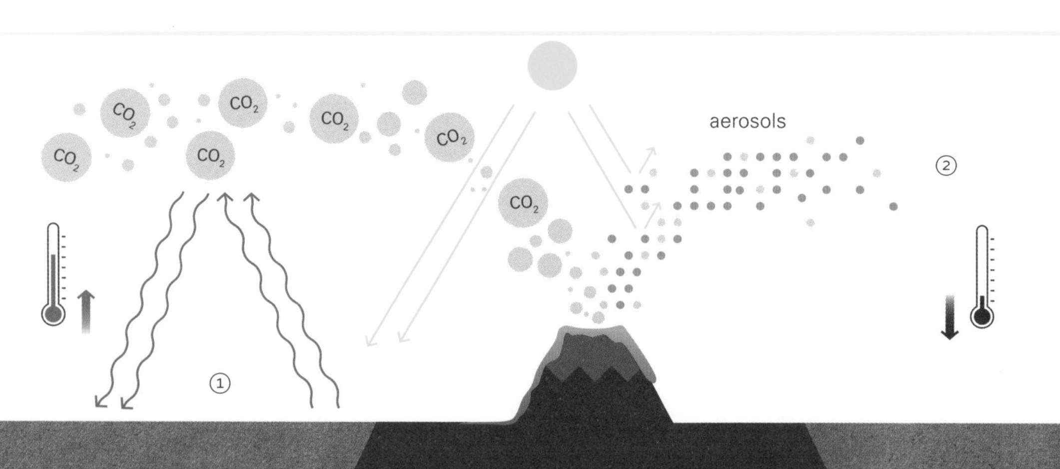

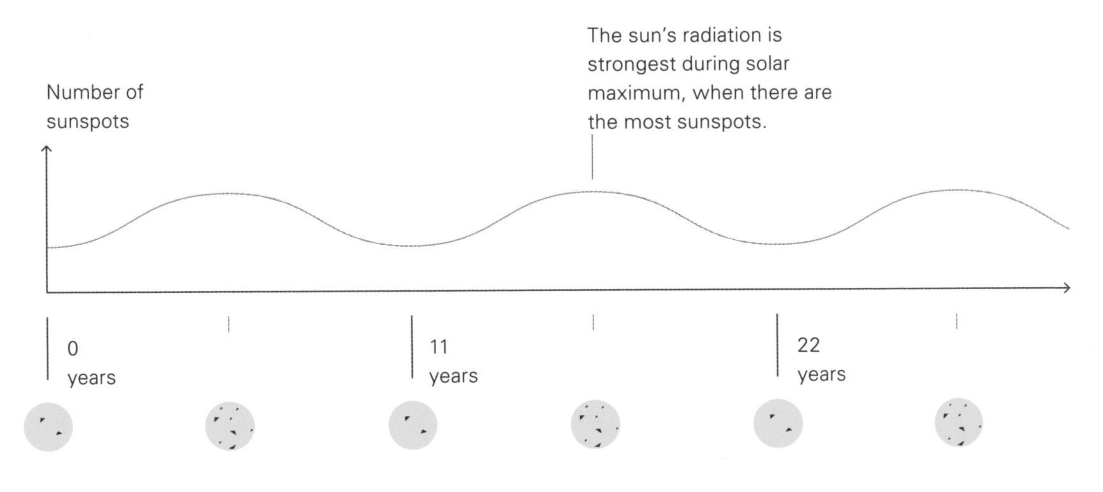

The 11-year sunspot cycle

Radiation from the sun is a prerequisite for life on Earth, and solar activity has played a role in some major changes to our climate in the past.[1,2] One measure of solar activity—that is, the intensity of the sun's radiation—is the presence of dark patches on the sun's surface, known as sunspots.[2,3] Their number varies on a roughly 11-year cycle—the sunspot cycle.[4,5] Besides the greater number of sunspots that can be observed at the peak of the cycle, which is known as "solar maximum," there is often also an increase in faculae (Latin for "little torches"), or bright solar flares. Solar flares represent an increase in solar radiation.[6] In the course of the sunspot cycle, solar activity periodically grows stronger and weaker, and this affects both global and regional temperatures on Earth.[7–10]

Clouds

Clouds scatter sunlight, which means that less radiation reaches the ground and Earth's surface is warmed to a lesser extent. Clouds also absorb the thermal radiation given off by Earth's surface, before releasing it again in all directions. This means that some of that energy is retained in Earth's system (which comprises the planet's land, sea, air, and living organisms). High-altitude, wispy cirrus clouds give Earth only partial protection from incoming solar radiation ❶. Formed of ice crystals, these clouds are also very cold, and as such they emit very little heat outward into space ❷. Cirrus clouds, therefore, generally have a warming effect. By contrast, low-lying clouds, which are usually much thicker, cool the average global climate because they scatter the bulk of incoming solar radiation back into space ❸. Besides this, these clouds are almost as warm as Earth's surface and therefore they release as much thermal radiation out into space as is released from the ground ❹.[1,2]

Under present conditions, clouds have a cooling effect overall.[3]

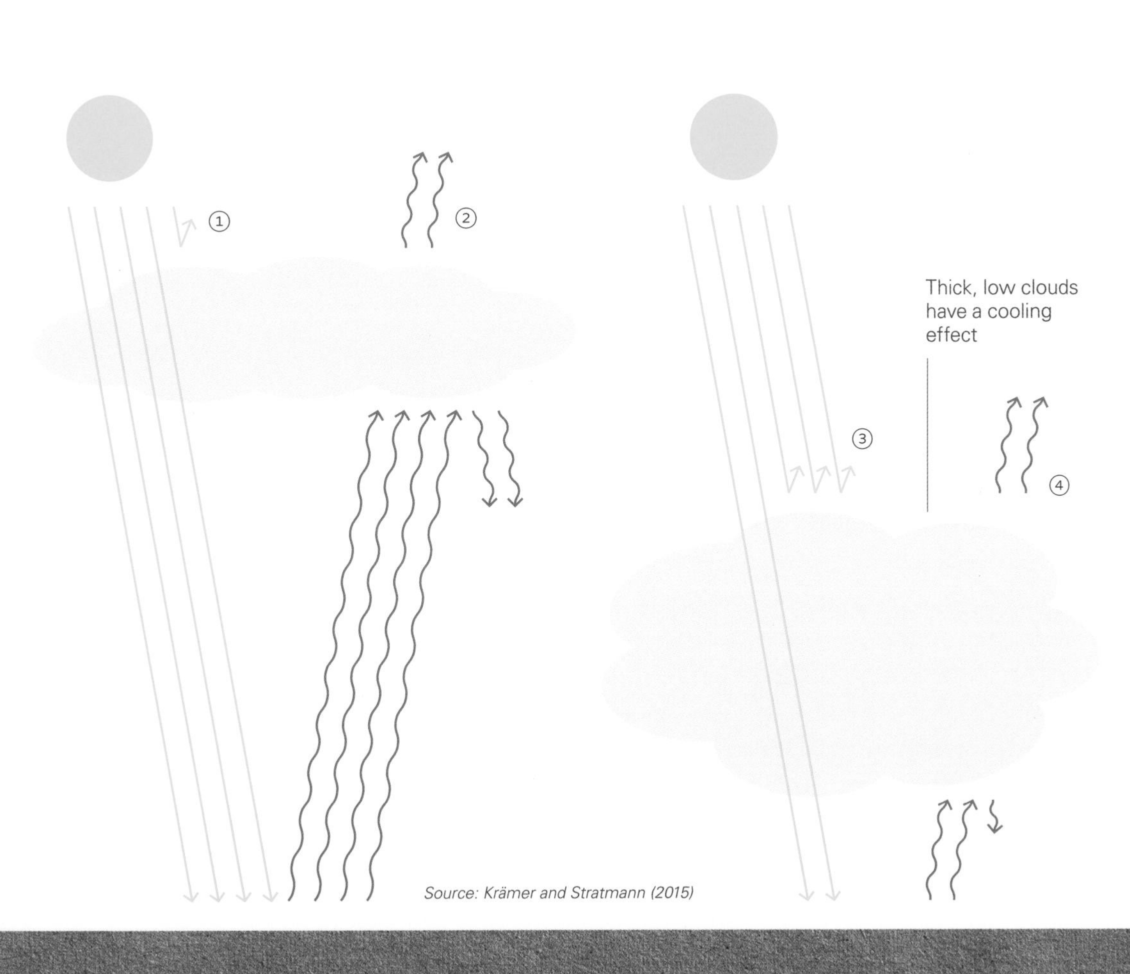

Source: Krämer and Stratmann (2015)

Ocean Circulation: the Global Conveyor Belt

The "global conveyor belt" is a simplified conceptual model of the complex system of currents that connects all of Earth's oceans.[1,2] Water circulates the globe, driven by winds sweeping over the surface, by tides, and by differences in seawater density (caused by variations in the temperature and salinity of the water).[3] This conveyor belt transports large quantities of heat that have a significant influence on the climate.[4,5] For example, if the Atlantic part of the "global conveyor belt" came to a complete standstill ❶, the air temperature in the northern hemisphere would drop on average by 1–2°C (1.8–3.6°F), and the temperature over the northern North Atlantic would decrease by as much as 8°C (14.4°F).[6]

— warm, surface currents
— cold, deep currents

Source: ACIA (2004)

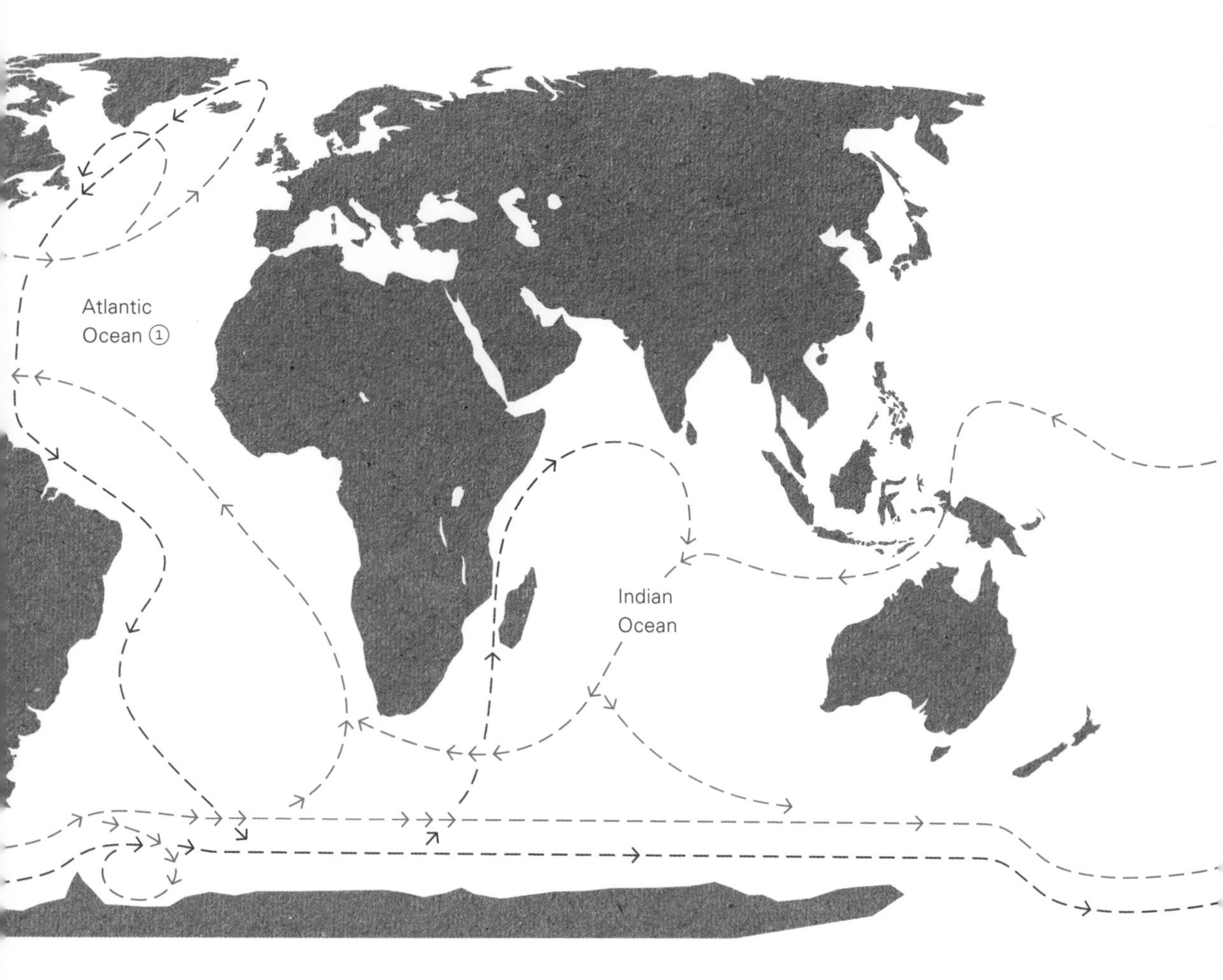
Atlantic
Ocean ①
Indian
Ocean

Climate History

The climate has changed constantly over the course of Earth's history, with regularly occurring extremes. One example is the Sturtian ice age, a glaciation triggered some 700 million years ago by a drop in solar radiation and a low concentration of carbon dioxide in the atmosphere ❶. With the cooling effect intensified by ice-albedo feedback (p. 52), vast swathes of Earth's surface froze over, turning our planet into a "Snowball Earth."[1]

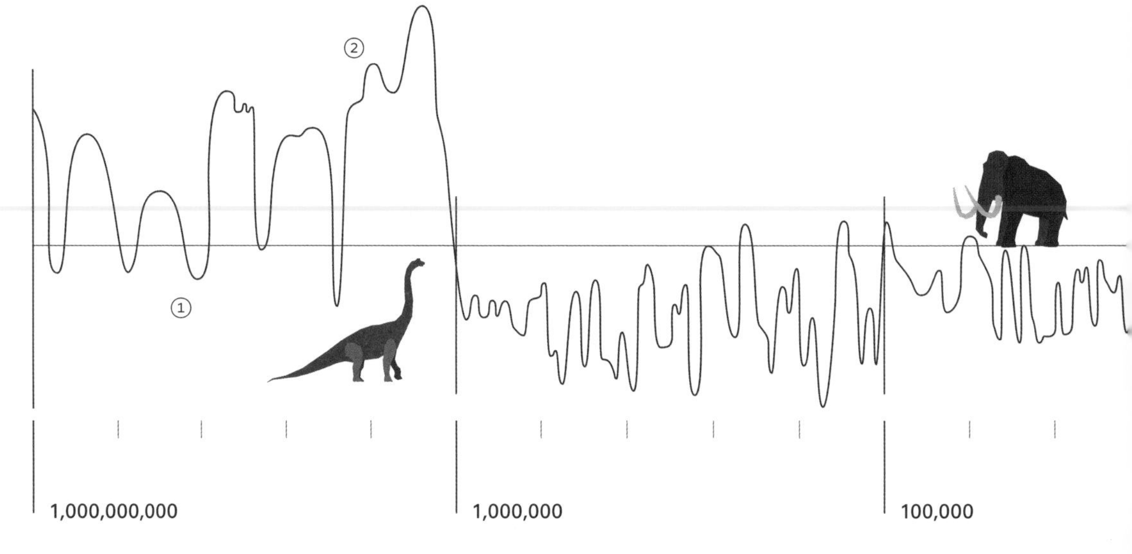

Schematic representation of the average surface air temperature (SAT) in the northern hemisphere

Around 250 million years ago, large quantities of carbon dioxide (CO_2) and methane (CH_4) were released into the atmosphere ❷.[2] This resulted in an intensification of the greenhouse effect and a sharp rise in temperature. The oceans became more acidic as they absorbed a portion of the CO_2 emissions (p. 68).[3,4,5] About 90 percent of all species living at that time became extinct.[6] Over the last 11,500 years, Earth's climate has been relatively stable ❸, a factor that has allowed for the development of modern civilization.[7]

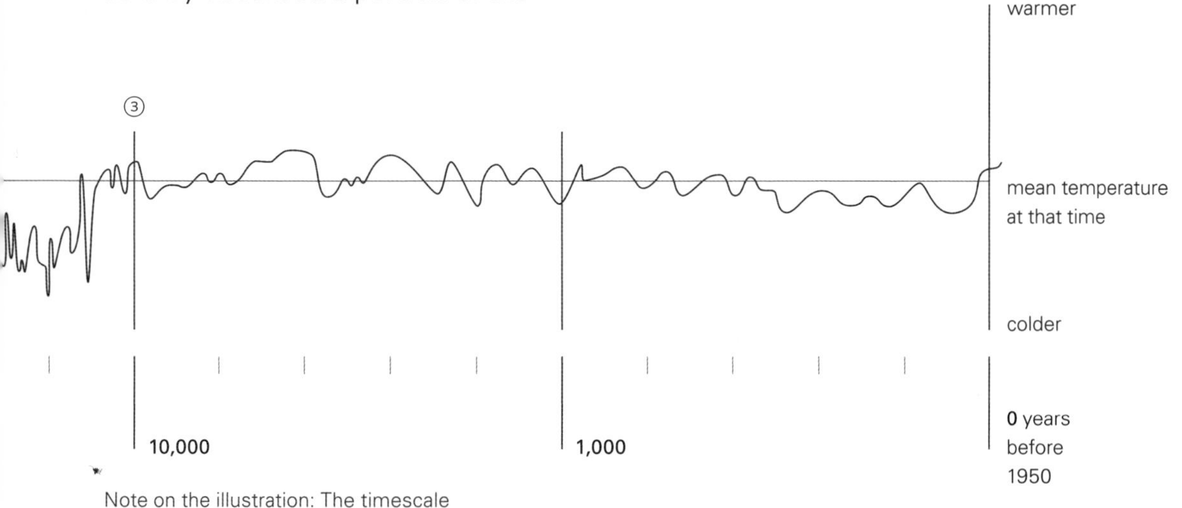

Note on the illustration: The timescale is consistent within each time period.

Source: Schönwiese (2013)

CHAPTER 2

POSSIBLE CAUSES OF CLIMATE CHANGE

The average global air temperature has risen in the last 150 years.[1] Besides human activity, in public debate about climate change we often hear about the sun and other factors as possible causes for this progressive increase.

Global Warming

In the northern hemisphere, the average ground-level air temperature stayed relatively constant in the thousand years leading up to the industrial revolution.[1] Since the end of the nineteenth century, we have observed a rise in the global temperature, a trend we today refer to as "global warming" or "climate change." From the time temperature records first began in 1880, up to 2016, the average global ground-level air temperature increased by more than 1°C (1.8°F).[2]

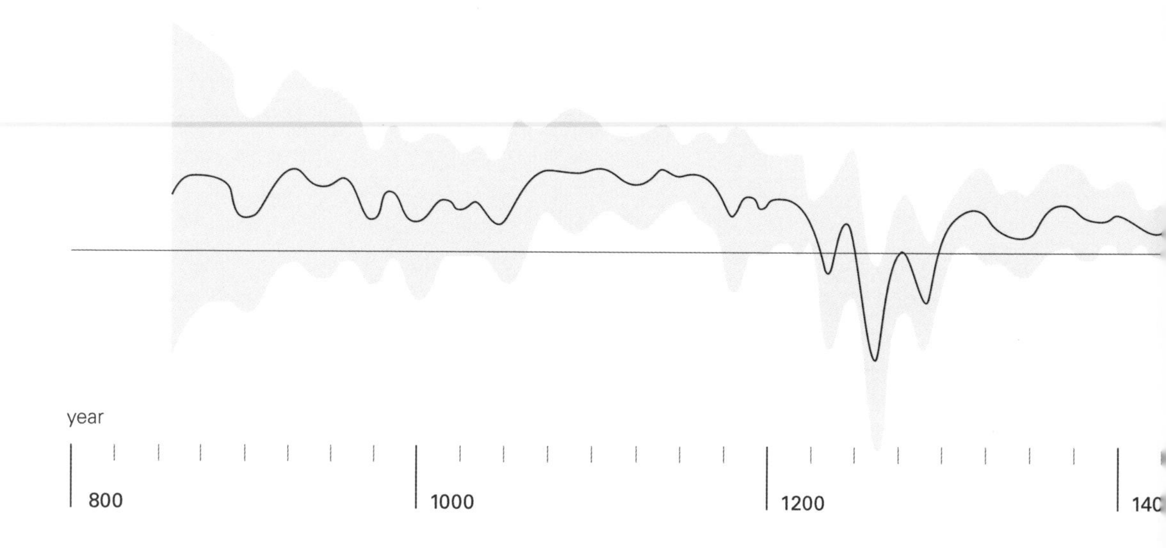

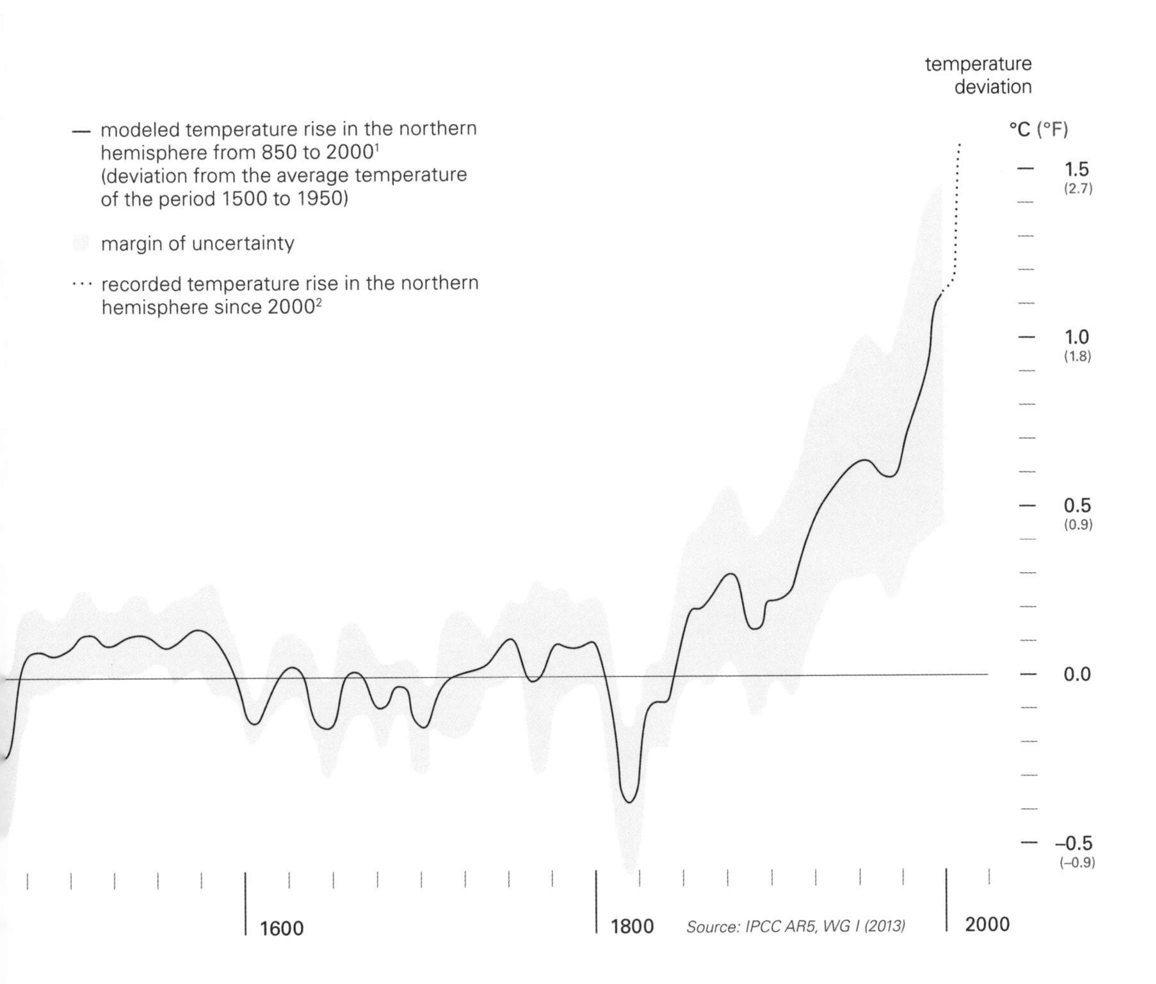

temperature
deviation
— modeled temperature rise in the northern hemisphere from 850 to 2000[1] (deviation from the average temperature of the period 1500 to 1950)
margin of uncertainty
··· recorded temperature rise in the northern hemisphere since 2000[2]
°C (°F)
1.5
(2.7)
1.0
(1.8)
0.5
(0.9)
0.0
–0.5
(–0.9)
1600
1800
Source: IPCC AR5, WG I (2013)
2000

The Ozone Layer

The ozone layer in the stratosphere absorbs a portion of solar radiation ❶, thereby protecting plants, animals, and people from the sun's harmful UV-C and UV-B radiation.[1,2] During the last sixty years, this layer has become thinner around the world. Over the Antarctic region, it has thinned by 50 percent or even more,[3] creating what we call a "hole in the ozone layer."[4] The reason for this thinning was the release of chlorofluorocarbons (CFCs) by humans.[5] CFCs are volatile substances used as coolants in refrigerators and air-conditioning units. They do not occur naturally in Earth's atmosphere, but, once released into it, they act as greenhouse gases.[6,7] Following the Montreal Protocol of 1987, 46 countries committed to reducing the emission of CFCs and other substances that harm the ozone

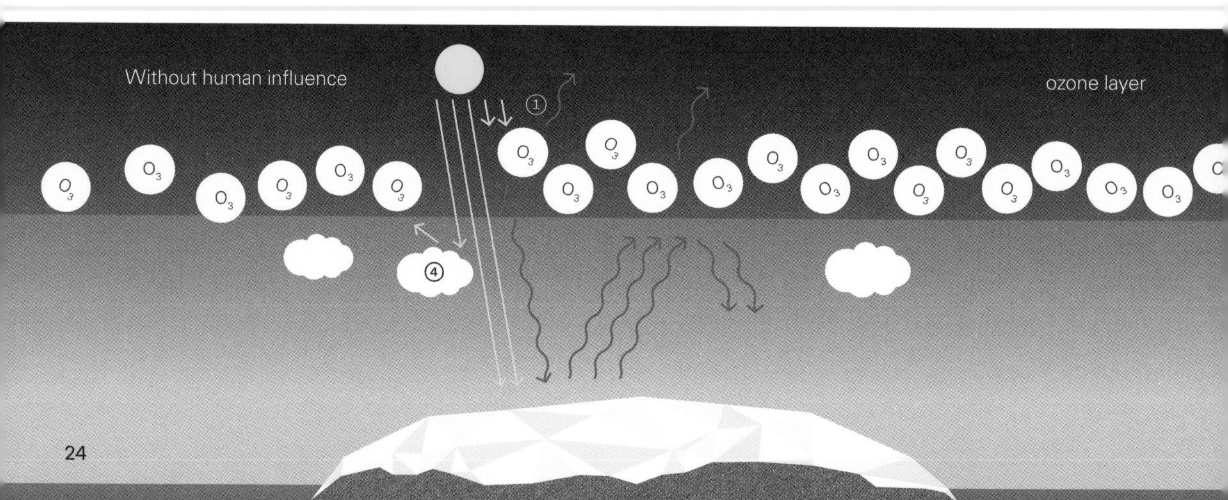

layer.[8] As a result, emissions of ozone-depleting substances have fallen since the 1990s, and the ozone layer will probably recover before the end of this century.[9] A thinner ozone layer ❷ allows more solar radiation to reach Earth's surface, but it also means the greenhouse effect is weakened ❸. The cooling effect is slightly greater than the warming effect overall.[10] However, this doesn't take into account the impact of the CFCs themselves on the greenhouse effect, or other feedback resulting from ozone depletion; for example, cloud formation is weakened ❹ and there are changes to atmospheric circulation (winds).[10,11]

The complexity of these feedback scenarios makes it difficult to assess whether the overall effect of ozone depletion will be warming or cooling.[9]

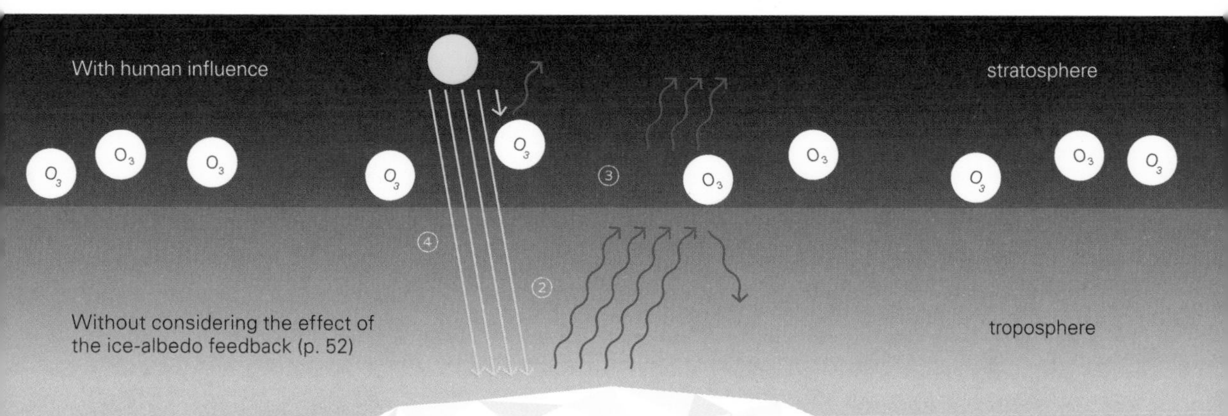

Aerosols

An aerosol is made up of fine matter (particles) suspended within a gas (generally, the air).[1] They form either when particles are directly emitted into the gas (primary aerosols) or when gases in the atmosphere are converted to solid or liquid particles (secondary aerosols).[2] They range in size from just a few nanometers to several tens of micrometers: up to 100,000 times smaller than the diameter of a human hair.[2,3] Naturally occurring aerosols can form when, for example, sea salt or desert dust gets whirled up into the air, when volcanic eruptions let off clouds of ash, or when vegetation releases airborne organic particles (for example, spores) ❶.[4–7] A significant proportion of global aerosols arise from human activities, such as slash-and-burn land clearance, industrial processes, and transportation ❷.[8–11]

Aerosols affect the climate by scattering sunlight so that less energy and heat reach Earth's surface ❸. On the other hand, aerosols like soot also absorb solar energy and release it into their surroundings, thus producing a warming effect ❹. Aerosols also influence cloud formation and change how much solar radiation they reflect ❺.[8] Human-made aerosols pollute the air, but, ironically, their overall effect on the climate is cooling and thus they attenuate global warming.[12,13]

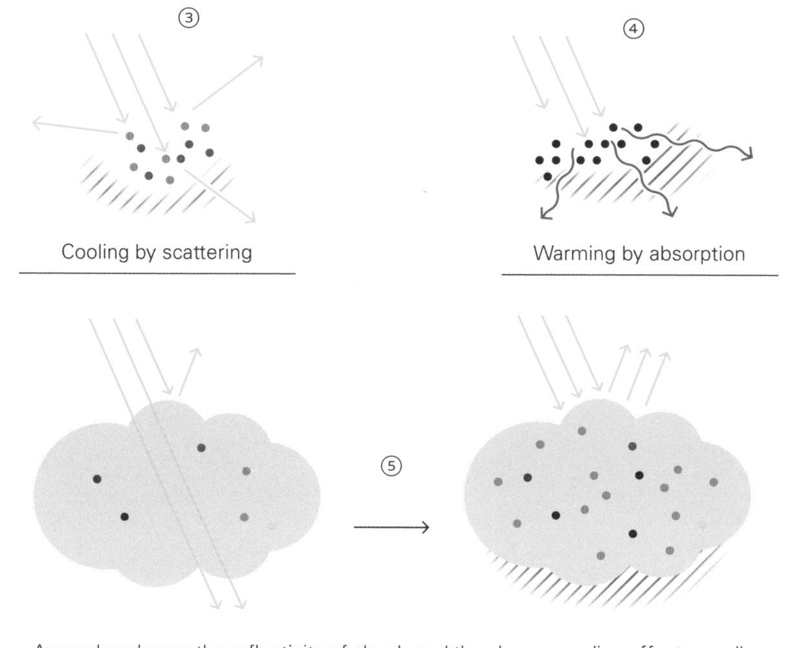

Aerosols enhance the reflectivity of clouds and thus have a cooling effect overall.

Source: IPCC AR5, WG 1 (2013)

Solar Activity

As we see in the graphic opposite, there is no direct correlation between solar activity ❶ and the sharp increase in average global temperature ❷ between 1880 and 2016.[1,2] It is thought that solar activity is responsible for only about 10 percent of the rise in temperature between 1905 and 2005.[3] As such, we can say that the contribution of solar activity to global warming is relatively small.[4] Since the 1980s, solar activity has in fact been in decline, whereas Earth's average air temperature has further increased.[3,5]

We cannot, therefore, ascribe the sharp rise in temperature since the beginning of industrialization to solar activity.

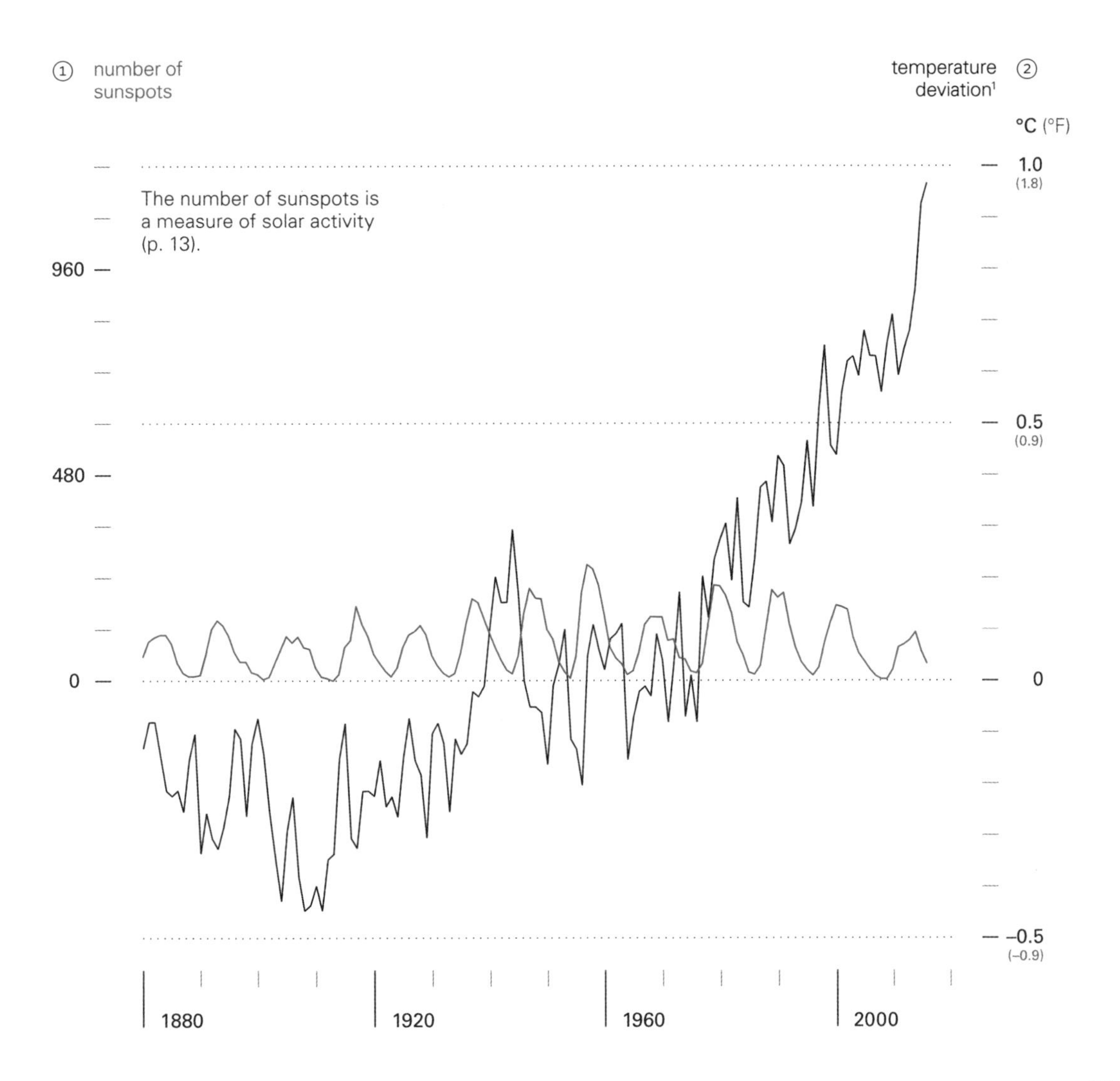
① number of sunspots
temperature deviation[1] ②
°C (°F)
1.0
(1.8)
0.5
(0.9)
0
–0.5
(–0.9)
960
480
0
The number of sunspots is a measure of solar activity (p. 13).
1880
1920
1960
2000

Human Activity

Since the start of the industrial era, there has been an increase not only in the average global air temperature ❶, but also in the concentration of carbon dioxide (CO_2) ❷ and other greenhouse gases in Earth's atmosphere.[1–12] Human activity, in particular the burning of fossil fuels, is the reason for this change.[13] Gases released by human activity are called anthropogenic greenhouse gases (p. 36) because, just like natural greenhouse gases, they prevent thermal radiation escaping from Earth back into space (p. 8).[14,15] As a result of these anthropogenic greenhouse gases, Earth's surface absorbs more thermal radiation than it otherwise would. This has contributed to the increase in surface air temperature over the last 150 years.[16,17]

The increase in temperature—from the start of industrialization until today—is therefore known as "human-made climate change."[18]

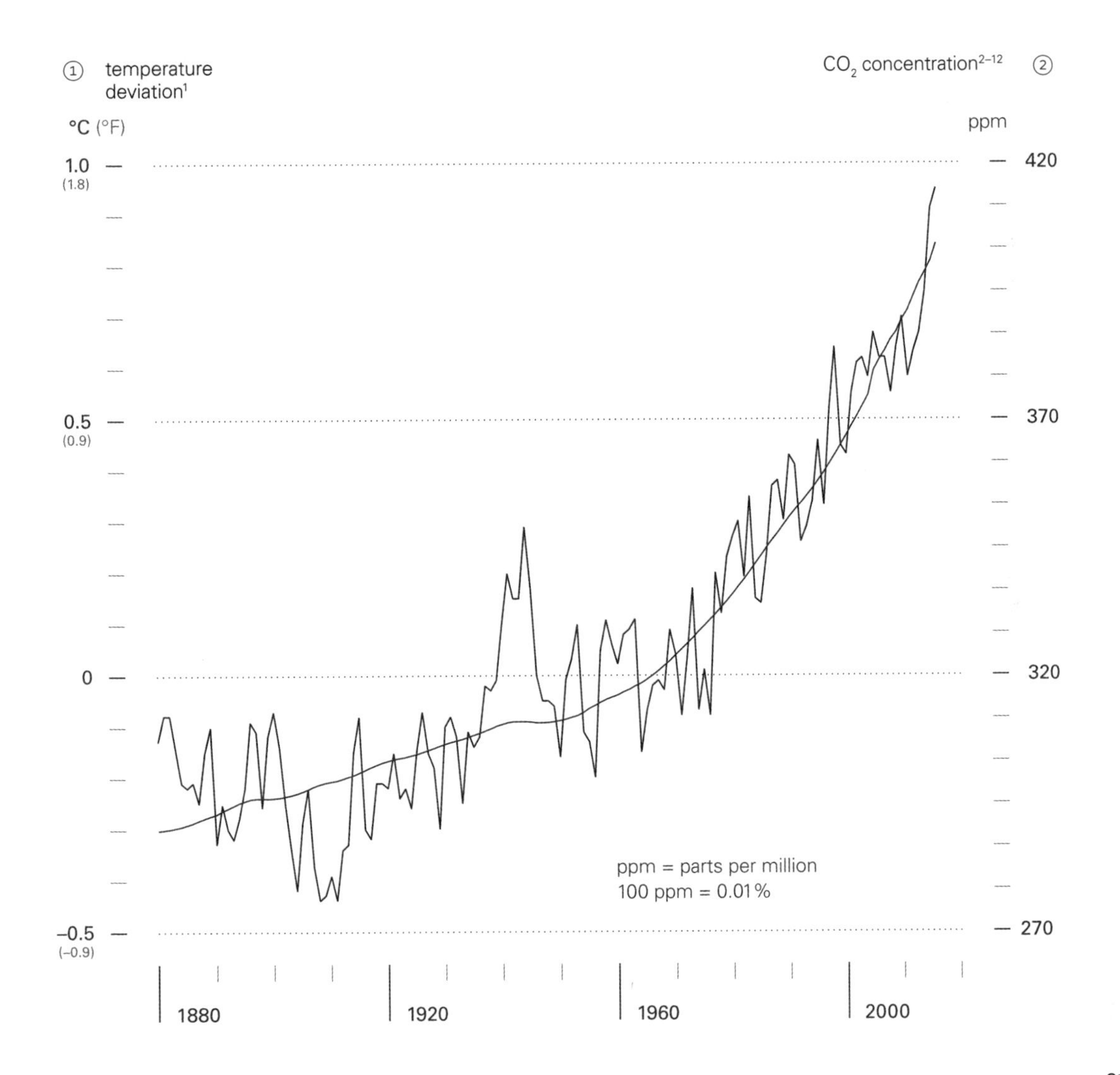

① temperature deviation[1]
CO_2 concentration[2–12] ②
°C (°F)
ppm
1.0
(1.8)
0.5
(0.9)
0
–0.5
(–0.9)
420
370
320
270
ppm = parts per million
100 ppm = 0.01 %
1880
1920
1960
2000

Temperature and Greenhouse Gases

Ice core samples allow us to reconstruct what the climate was like several hundred thousand years ago.[1] Deep holes are drilled into ice sheets to extract columns, or cores, of ice.[2] By analyzing the gases and solids trapped in the various layers of ice, scientists can draw conclusions about past temperatures, greenhouse gas concentrations and volcanic eruptions.[3,4]

The graph on the right shows that ice ages and warm periods have alternated over the last 800,000 years ❶. Though there are various reasons for the natural variations in climate (such as a change in Earth's orbit and rotational axis), a correlation can be clearly seen between temperature and the presence of the greenhouse gases carbon dioxide (CO_2) and methane (CH_4) ❷.[5–8] In addition, recent studies confirm that changes in temperature and the concentration of greenhouse gases over the last 20,000 years have occurred in parallel, suggesting an interrelationship.[9,10] It is also clear that today's concentrations of greenhouse gases are significantly higher than at any other point in the last 800,000 years ❸.[11,12]

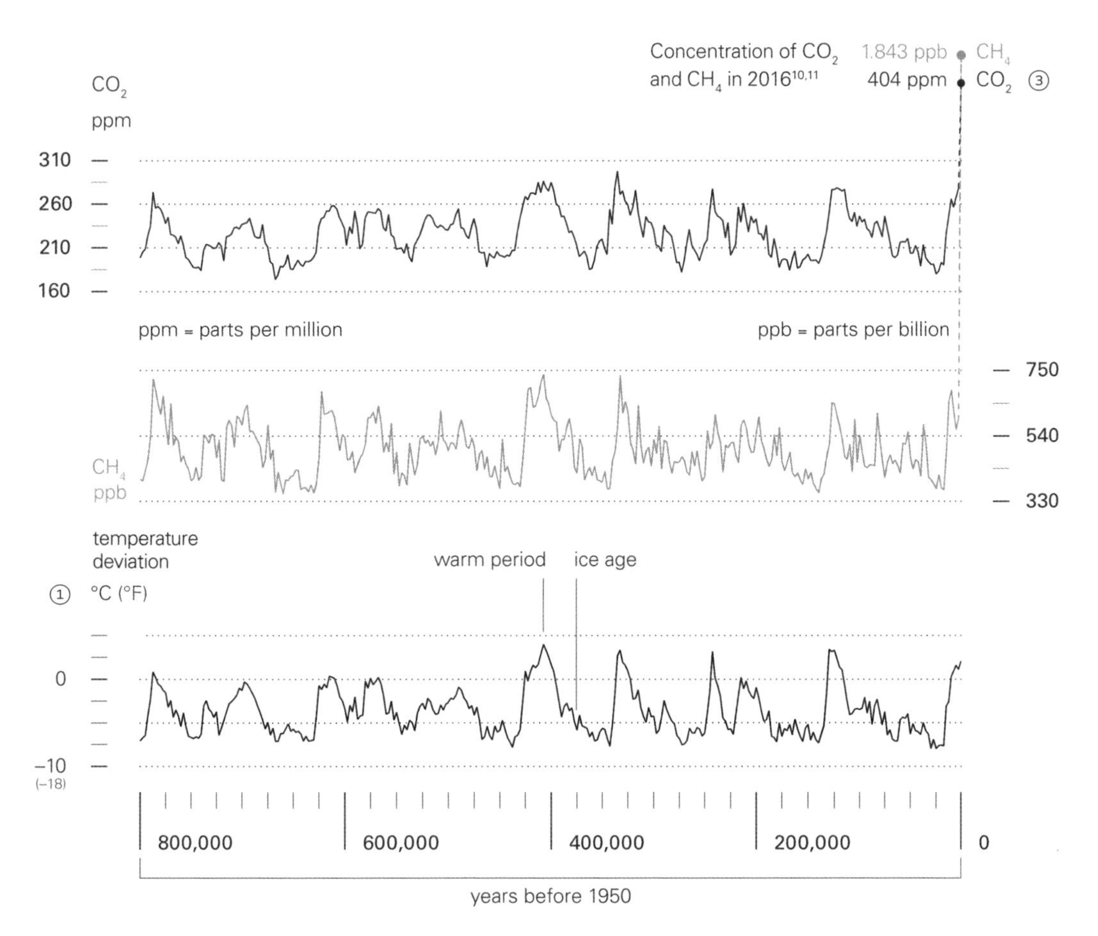

② The temperature in the Antarctic and greenhouse gas concentrations over the past 800,000 years show a close correlation.[5,6,7]

Factors Contributing to Global Warming

A number of studies have investigated the role played by natural factors (for example, solar radiation and volcanoes) and human activity in accelerating global warming.[1–7] These studies have revealed that the increase in global air temperature since the beginning of industrialization cannot be explained without reference to human activity.

Between 1870 and 2010 the average global ground-level air temperature was repeatedly cooled over short periods by volcanic eruptions (p. 12) ❶.[8] Fluctuations in solar activity (p. 28) have likewise had only a minor impact on the climate ❷.[9]

Other relatively short-term changes in temperature can be explained by "internal variability" within the Earth system ❸,[10,11] including the interactions between oceanic and atmospheric circulation (winds). These natural fluctuations occur independently of incoming solar radiation and the thermal radiation emitted by Earth. As illustrated in the graph, the Intergovernmental Panel on Climate Change (IPCC) estimates that the warming of 0.7°C (1.3°F) from 1951 to 2010 is largely human-made ❹.[12] In contrast, the contribution of natural factors to the increase in temperature is estimated as just ±0.1°C (0.2°F).[12]

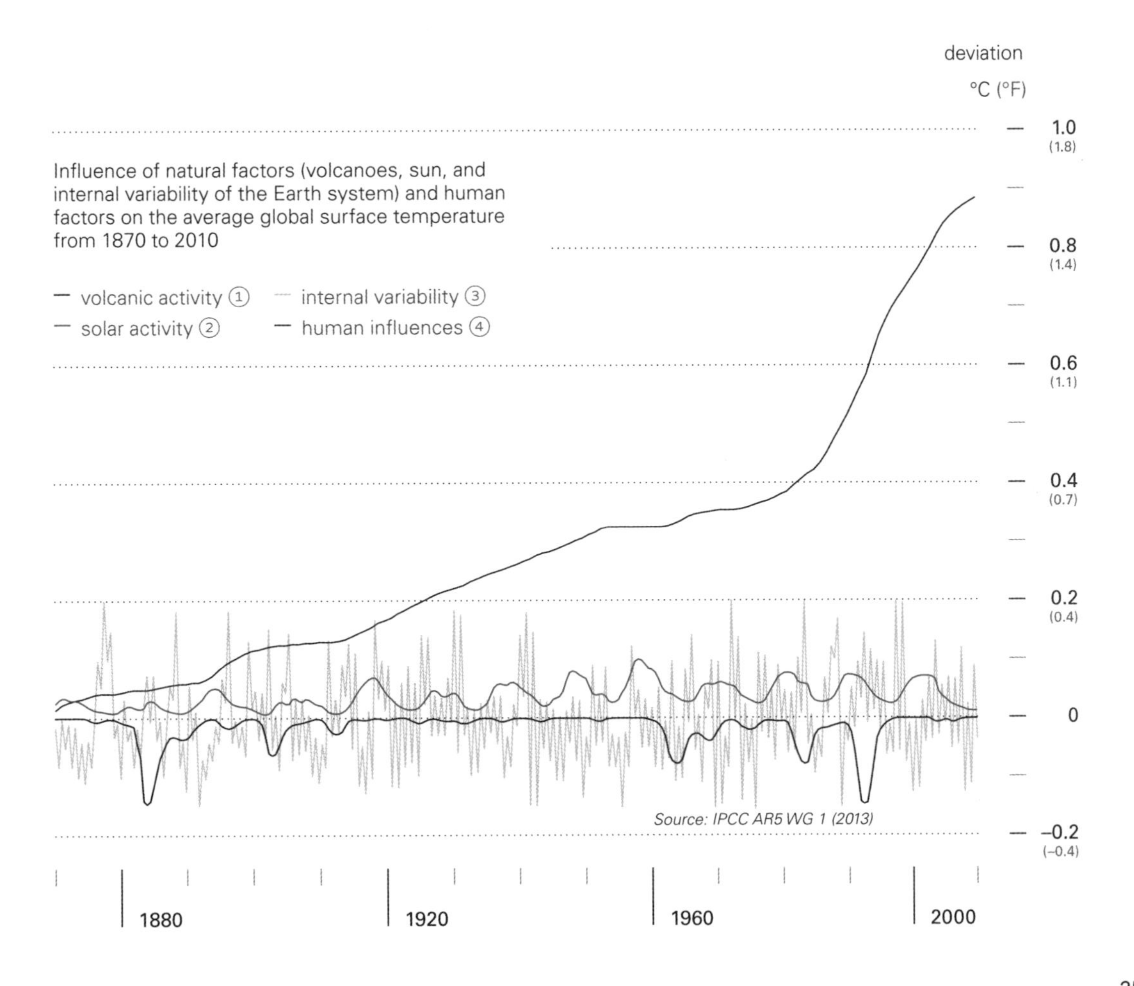
deviation
°C (°F)
1.0
(1.8)
0.8
(1.4)
0.6
(1.1)
0.4
(0.7)
0.2
(0.4)
0
–0.2
(–0.4)
Influence of natural factors (volcanoes, sun, and internal variability of the Earth system) and human factors on the average global surface temperature from 1870 to 2010
volcanic activity ①
solar activity ②
internal variability ③
human influences ④
Source: IPCC AR5 WG 1 (2013)
1880
1920
1960
2000

Anthropogenic Greenhouse Gases

Greenhouse gases emitted as a result of human activities are referred to as anthropogenic, or human-made, greenhouse gases. How much CO_2, CH_4 and N_2O emissions contribute to the anthropogenic greenhouse effect depends on their concentration in Earth's atmosphere (quantity) and on their greenhouse-gas potential (impact).

The proportion of greenhouse gases—carbon dioxide (CO_2), methane (CH_4) and nitrous oxide (N_2O)—in Earth's atmosphere is still relatively low in comparison with oxygen (21 percent) and nitrogen (78 percent).[1] However, since the start of industrialization, there has been a sharp increase in concentrations of these gases, as a result of rising emissions from human activity ❶.[2] Just like natural greenhouse gases, they prevent the direct escape of thermal radiation from Earth into space (p. 8) and thus contribute to global warming.[3,4] The gases also exert a long-term influence on the climate because of their long atmospheric lifetime ❷. In other words, it takes a long time before they are removed from the atmosphere or destroyed by chemical or physical processes.[5]

The impact of a particular greenhouse gas depends on its Global Warming Potential (GWP) ❸.[12] This describes the potential of the gas to trap heat in Earth's atmosphere, compared to the same amount of CO_2 in a given period of time (usually 100 years).[13] For example, a GWP of 28 for methane means that CH_4 emitted today will warm the climate over the next 100 years 28 times as much as the same amount of CO_2 would.[11]

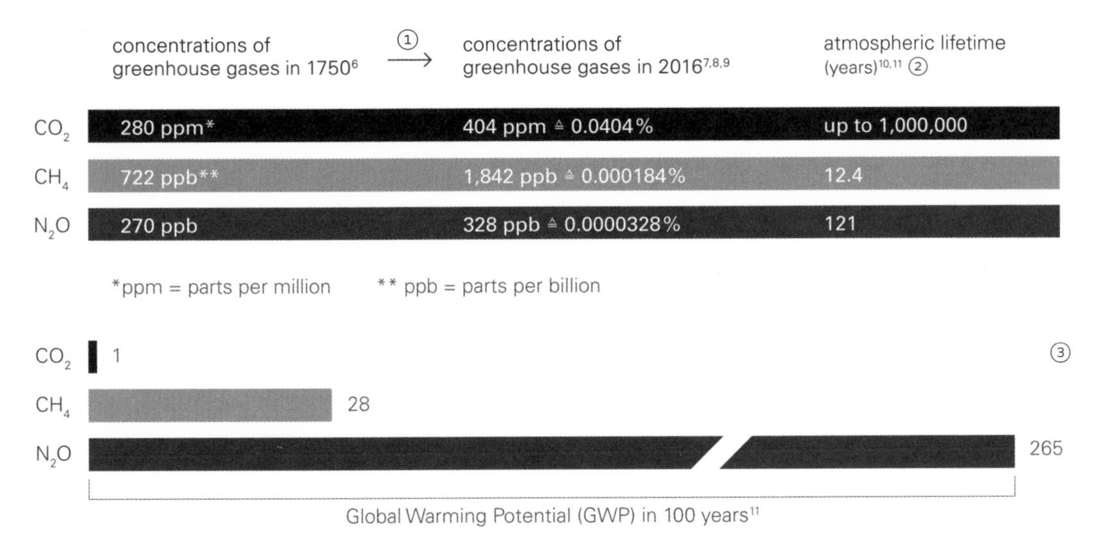

Even though the GWP of CO_2 is much smaller than that of CH_4 and N_2O, emissions of CO_2 make the biggest contribution (76 percent) to the overall anthropogenic greenhouse effect ❹, because the quantities of CO_2 emitted by human activities are far greater than those of CH_4 or N_2O.[14]

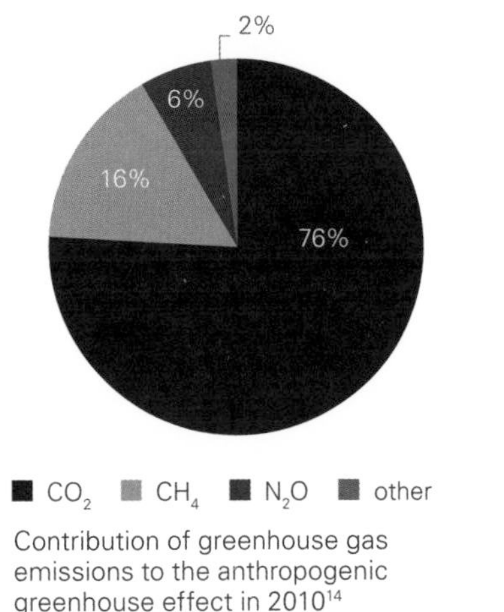

Contribution of greenhouse gas emissions to the anthropogenic greenhouse effect in 2010[14]

Changes to the Carbon Cycle

Oceans, soils, and vegetation release carbon dioxide (CO_2) into the atmosphere. They also absorb CO_2 from the atmosphere, and therefore form part of the natural carbon cycle.[1,2]

In the past ten years, humans have released 39 gigatons (43 billion US tons) of CO_2 each year.[3] About 28 percent of it is captured by terrestrial vegetation and soils, and about 22 percent is absorbed by the oceans, but the rest (44 percent) remains in the atmosphere.[3] The fact that, in this relatively short time period, human activity has released so much more CO_2 into the atmosphere has knocked the natural carbon cycle off-kilter. Much of this carbon was underground for millions of years in the form of coal, natural gas, or crude oil.[4] One consequence of this large and sudden release of CO_2 is that the oceans have become more acidic (p. 68); another is that CO_2 concentrations in Earth's atmosphere are much higher than they have ever been in the last 800,000 years (p. 32).[5–8]

Average annual CO_2 emissions by humans between 2007 and 2016, and net absorption of those emissions by soils and vegetation, the oceans, and the atmosphere[3]

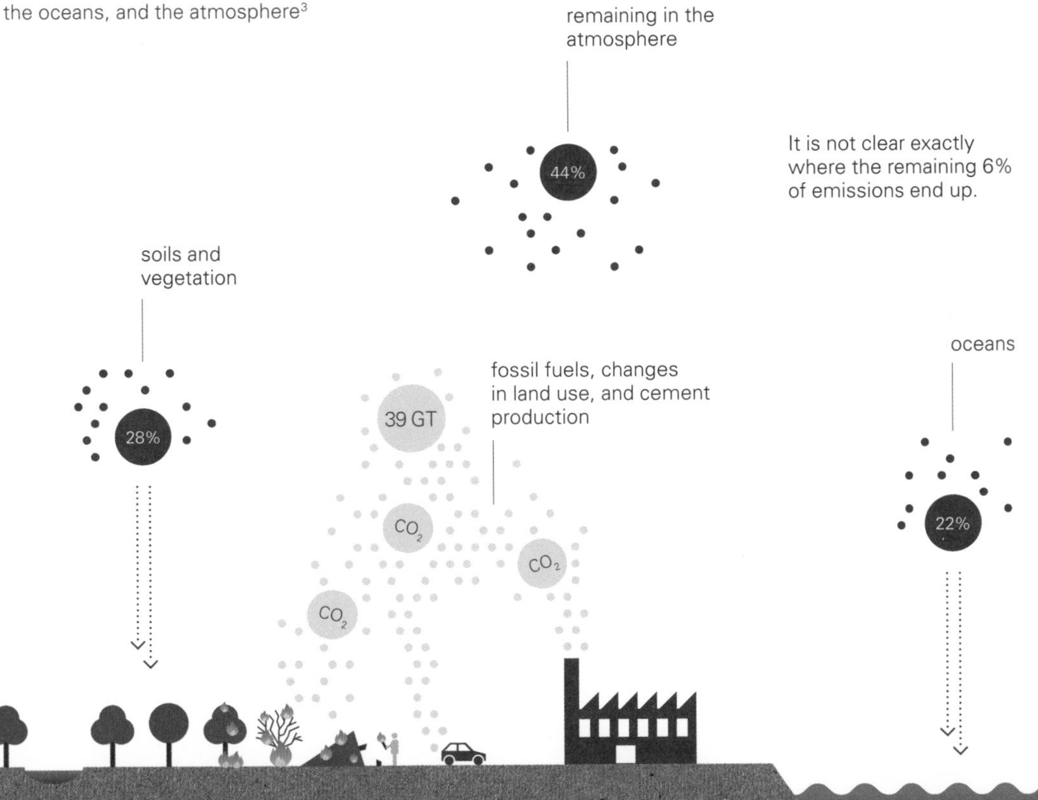

Carbon Dioxide Emissions

The burning of fossil fuels (coal, petroleum, and natural gas) to generate energy accounted for approximately 85 percent of global CO_2 emissions in 2014. Cement production was responsible for 5 percent, and changes in land use for 10 percent.[1] This shows that the burning of fossil fuels is primarily responsible for the increase in the concentration of CO_2 in Earth's atmosphere. The diagram on the right shows how the energy derived from fossil fuels is used.[2] Below, we see that coal, at 44 percent, accounts for the largest share of emissions generated by burning of fossil fuels.[1] Another source of CO_2 emissions is deforestation (i.e., changes in land use), which occurred prior to the industrial age with the large-scale felling of trees in North America and Europe several hundred years ago.[3,4] Today, it is mainly tropical rainforest that is being cut down and cleared to build roads, develop pastureland, produce timber, or to plant crops, such as oil palms, bananas, soy, and coffee, which are then sold to other countries.[5–9] This, together with natural causes (such as forest fires), has meant that in the period from 2000 to 2009 forested areas have been lost at an average rate of 35 soccer fields per minute.[10,11]

Contribution of coal, oil, and gas to fossil fuel emissions[1]

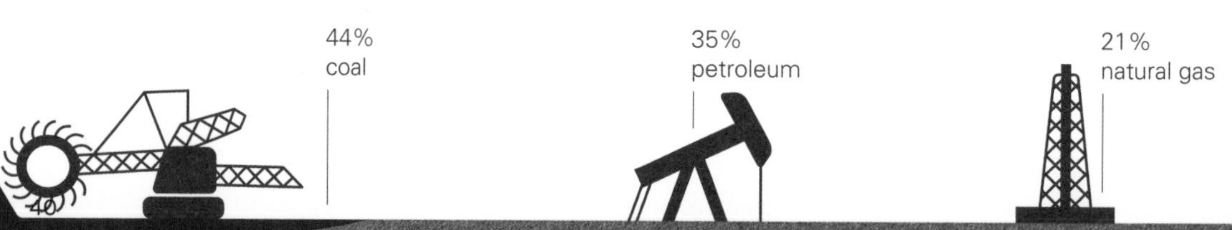

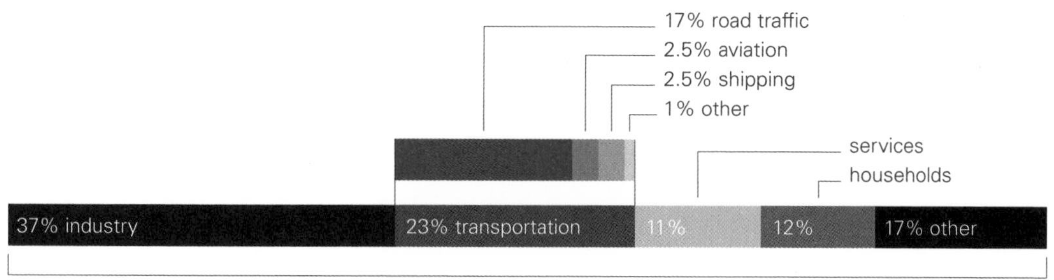

CO_2 emissions by sector in 2014[2]

Slash-and-burn clearance of tropical rainforest releases the carbon (C) stored in trees and peat soils in the form of CO_2.[12,13]

Methane and Nitrous Oxide Emissions

In the period from 2000 to 2009, 29 percent of global human-made methane (CH_4) emissions came from fossil fuel extraction.[1,2,3] Nearly as much methane is generated by livestock farming, principally from the digestive emissions of cattle.[4] Almost another quarter is released by decomposing waste in landfill sites.[5] Similar decomposition processes are triggered when rice fields are flooded (wet rice cultivation), which also releases methane into the atmosphere.[6] The remaining emissions come from burning biomass (for example, wildfires and bush fires), and manufacturing biofuel, such as that derived from oil palms.[1,7] Agriculture is by far the largest contributor to nitrous oxide (N_2O)

Global, human-made methane emissions[1]

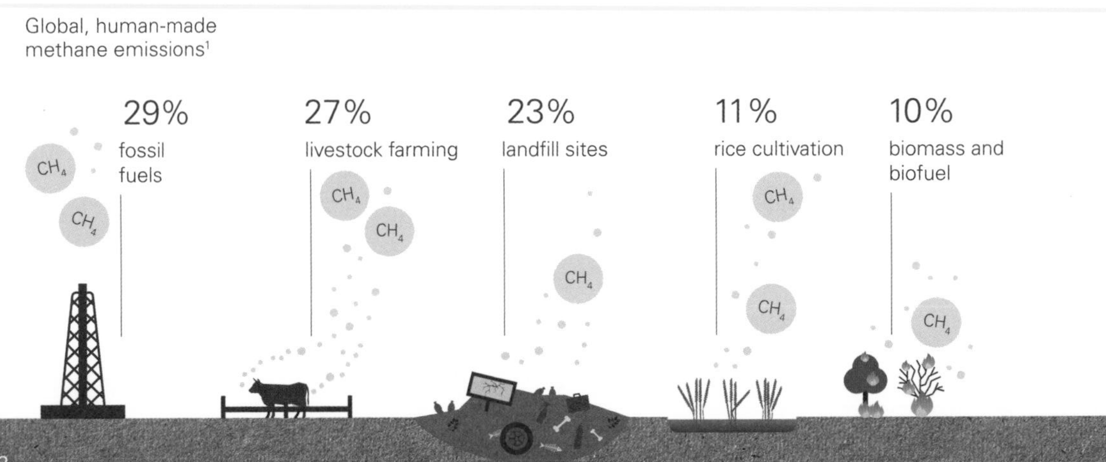

emissions, with a share of 59 percent.[1] The fertilizers used in farming contain nitrogen compounds that are partly broken down by bacteria in the soil. This causes nitrous oxide to be released into the atmosphere.[8] Livestock also produce N_2O through their excretions.[9] The combustion of biomass and biofuel—as with the burning of fossil fuels—accounts for only a small proportion (10 percent) of total human-made N_2O emissions.[1] Rivers also release nitrous oxide, since nitrogen compounds are washed into waterways (for example, through sewage and fertilizers used in farming), where they are then broken down by bacteria.[10,11] The remaining emissions come from other sources, such as human excrement.[1]

Global, human-made nitrous oxide emissions[1]

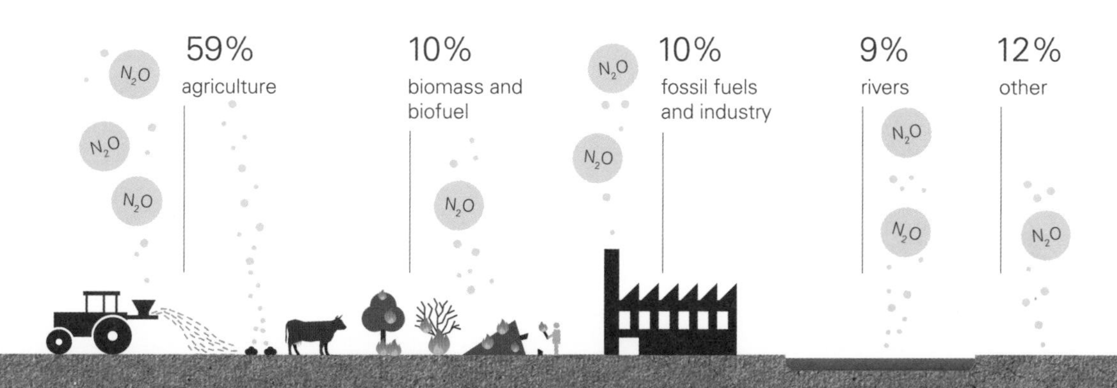

Carbon Dioxide Emissions by Country

In 2015, China was by far the biggest emitter of carbon dioxide from the burning of fossil fuels, ahead of the USA and the European Union ❶.[1,2]

However, a different picture emerges if we look at historical emissions, which still contribute to global warming today because of the long atmospheric lifetime (p. 36) of CO_2. Between 1918 and 2012, the EU and the USA emitted much more CO_2 than China ❷.[3] The USA and the EU are therefore mainly responsible for the rise in global average temperature since the start of industrialization. In order to compare the emissions from various countries, a country's CO_2 emissions are divided by the respective population size, giving the CO_2 emissions per capita ❸. However, emissions resulting from the manufacture of goods are generally allocated to the country where they are produced (the "producer pays" principle). If per capita emissions were calculated according to the "consumer pays" principle—i.e., by allocating emissions to the country in which the goods are eventually consumed—we would see a reduction in the per capita emissions of China, India, and Russia, and a rise in the emissions of European countries and the USA ❹.[4]

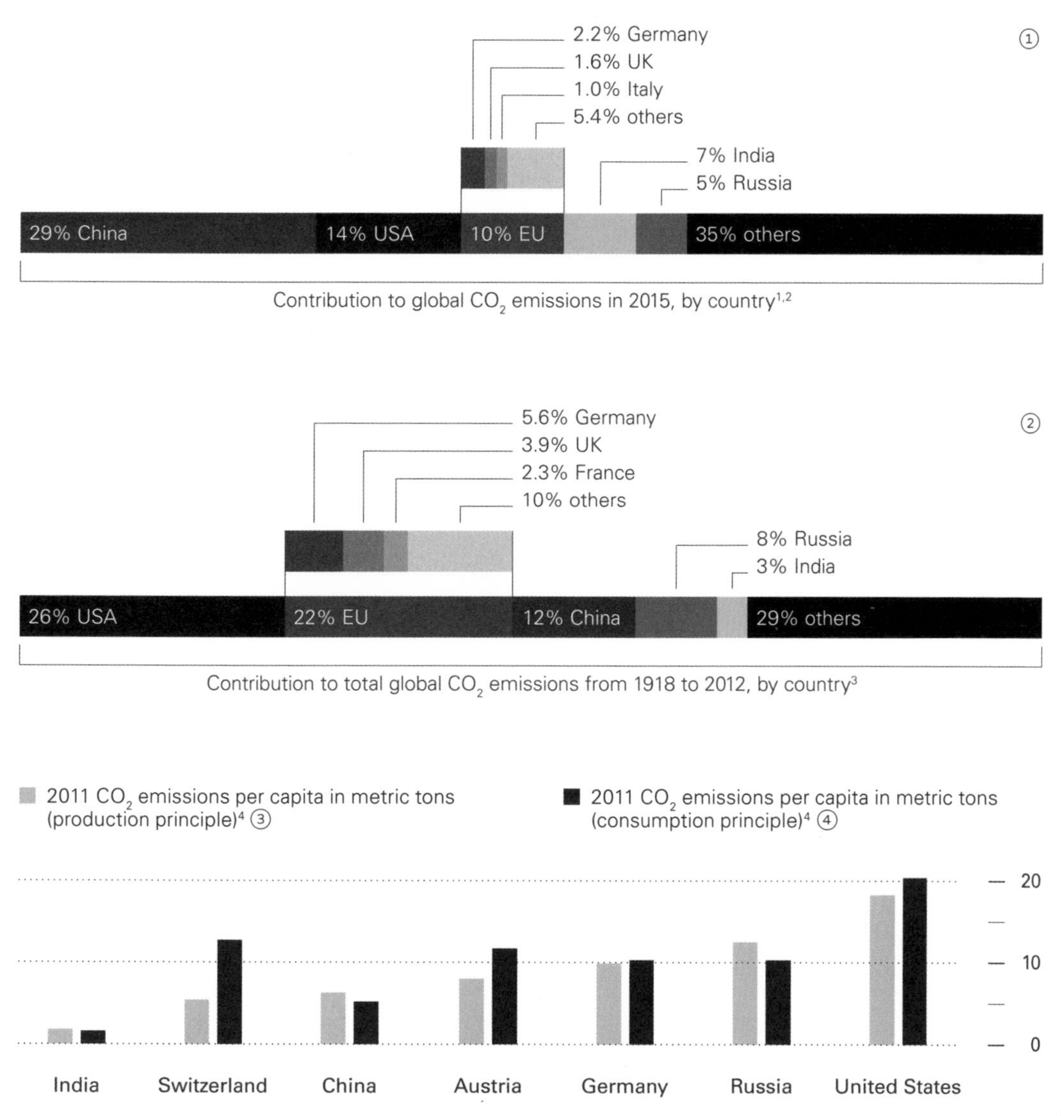
2.2% Germany
1.6% UK
1.0% Italy
5.4% others
7% India
5% Russia
29% China
14% USA
10% EU
35% others
①
Contribution to global CO_2 emissions in 2015, by country[1,2]
5.6% Germany
3.9% UK
2.3% France
10% others
8% Russia
3% India
26% USA
22% EU
12% China
29% others
②
Contribution to total global CO_2 emissions from 1918 to 2012, by country[3]
2011 CO_2 emissions per capita in metric tons (production principle)[4] ③
2011 CO_2 emissions per capita in metric tons (consumption principle)[4] ④
20
10
0
India
Switzerland
China
Austria
Germany
Russia
United States

Other Evidence of Human Influence

Satellites have allowed scientists to measure how much thermal radiation from Earth's surface is released back into space. These measurements show that, since 1970, less and less of the radiation that greenhouse gases can absorb has been released back into space. This is because the rise in greenhouse gas concentrations increasingly prevents thermal radiation from leaving the atmosphere ❶.[1] Other observations show that more thermal radiation is being reflected back onto Earth's surface ❷.[2,3] As such, the troposphere (lower atmosphere) has been warming, while the stratosphere (the next atmospheric layer up) has demonstrably cooled ❸.[4]

Without human influence

With human influence

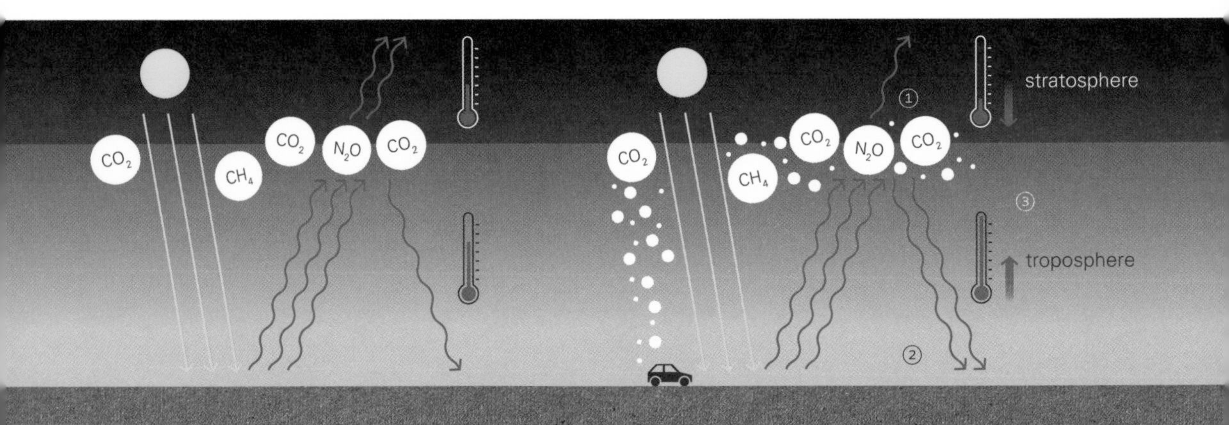

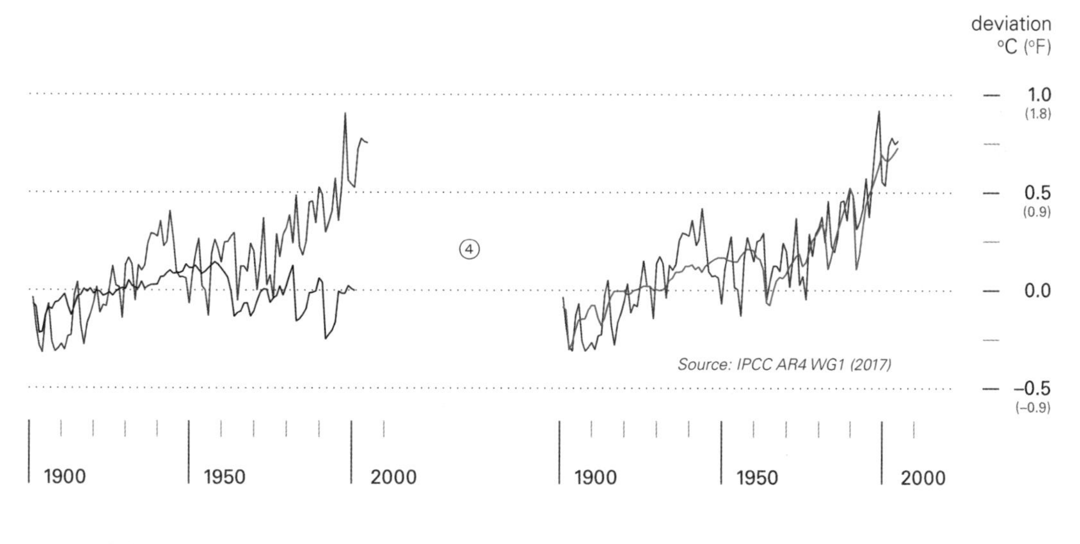

The observational evidence demonstrates and confirms the theory that climate change today is a human-made phenomenon. Furthermore, climate model simulations have shown that the recorded rise in temperature cannot be explained without the influence of humans ❹.[5]

CHAPTER 3

THE CRYOSPHERE

The cryosphere is the term used to describe all parts of Earth covered by water in its frozen form—in other words, all areas that are covered by snow and ice, glaciers, and permafrost soils.[1]

All surfaces reflect a certain proportion of incoming radiation. The measure of how much radiation is reflected is called the "albedo."[2]

Permafrost refers to ground that remains at a temperature of 0°C (32°F) or colder over a period of at least two consecutive years.[3]

Land ice is formed from compacted snow.[2]

An ice shelf is a floating mass of land ice that has flowed from the land into the sea.[4]

Sea ice forms when seawater freezes.[2]

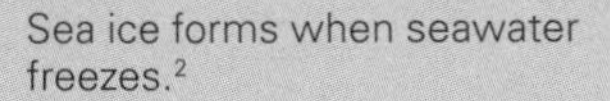

The Arctic

Sea ice is formed when seawater freezes. It has a lower density than seawater, which allows it to float on the surface.[1] It can become up to several meters thick. Only roughly 12 percent of sea ice is visible above the water.[2,3]

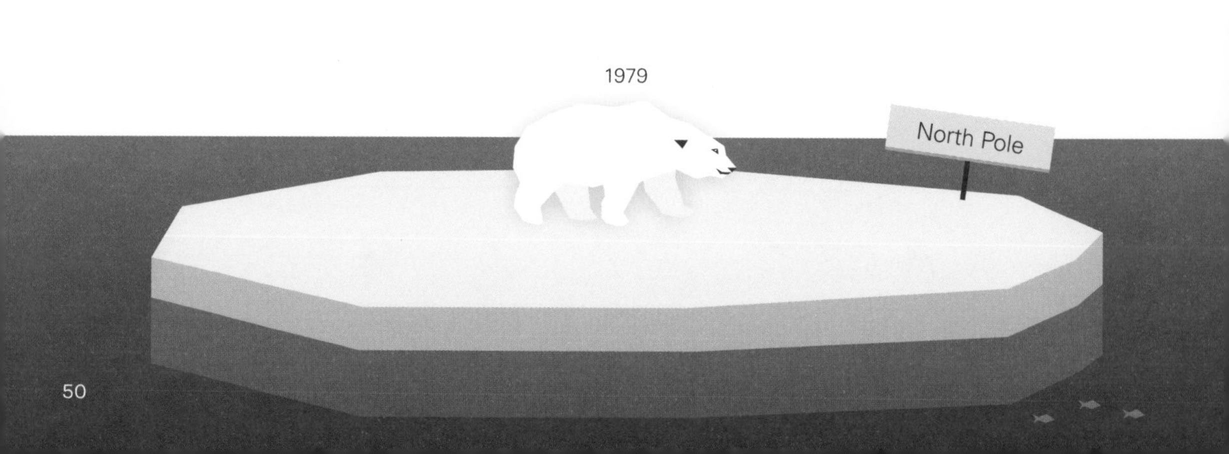

The consequences of climate change are particularly clear in the Arctic, because the air temperature there is increasing significantly faster than the average air temperature of Earth as a whole.[4,5] From 1979 to 2016, the minimum area covered by sea ice in the Arctic (measured in September each year) fell by approximately 43 percent. This corresponds to an annual decrease of an area larger than the state of Maine.[6,7] In the same period, the thickness of the ice cover also shrank, meaning that the total volume fell by around 77 percent. This is a clear indicator of an overall loss of sea ice in the Arctic. To picture the quantity of ice that has been lost, imagine all of California covered by a sheet of ice about 93 feet (28 m) thick.[8,9,10]

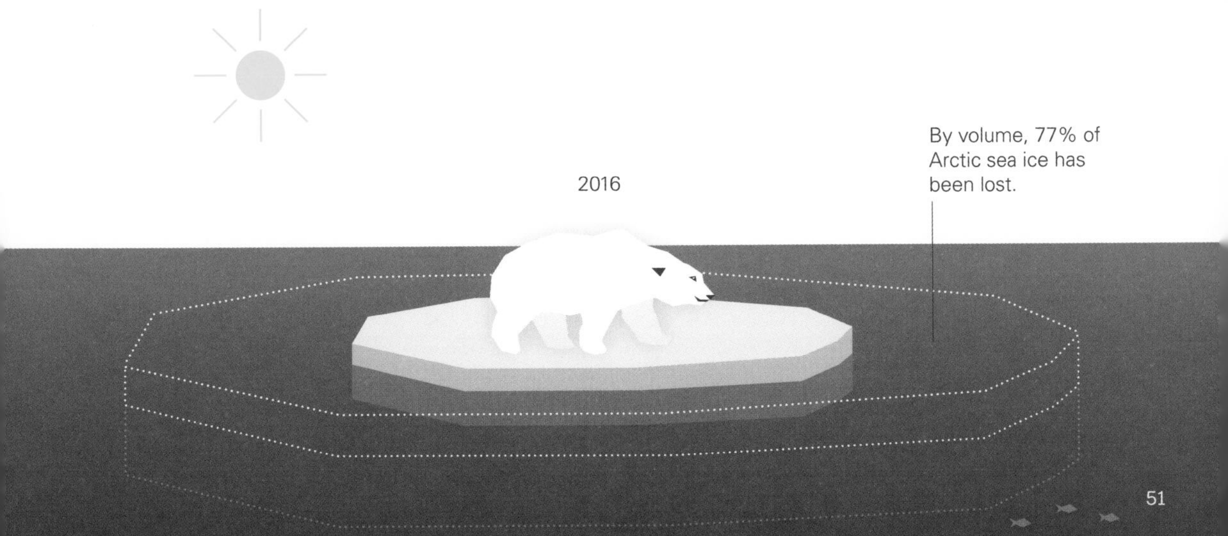

Ice-Albedo Feedback

Different surfaces reflect different amounts of incoming solar radiation back out to space.[1,2] For example, snow reflects more radiation than an area covered by forest. The proportion of reflected radiation a given surface can reflect is called "albedo."[2]

Snow and ice reflect a huge proportion of incoming solar radiation back into space (high albedo). If an area covered with snow or ice melts because of higher temperatures, the generally darker area beneath—for example, water or rock—will then become exposed, reflect much less radiation (low albedo), and instead absorb heat and warm up further.[3] This causes further warming of Earth, resulting in even more snow and ice melt, and even more warming again. This self-reinforcing, accelerating

Starting point

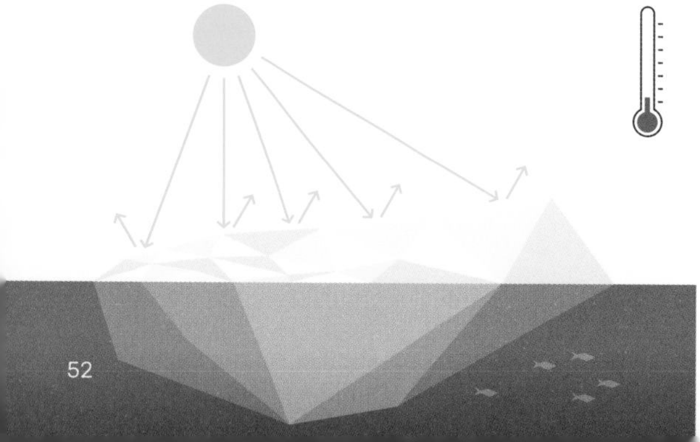

The ice melts because of higher temperatures.

process is known as the ice-albedo feedback.[4]

The ice-albedo feedback effect plays a particularly important role in the Arctic. Since an increasing surface area of sea ice now melts during each Arctic summer, the ocean absorbs a much greater amount of heat than would be the case if the ice cover were still in place. Since the ocean is correspondingly warmer, the ice now melts not only because of solar radiation, but also, increasingly, because of the warmer seawater that surrounds it, further intensifying the melting effect.[5]

Melting snow and ice contribute to global warming.

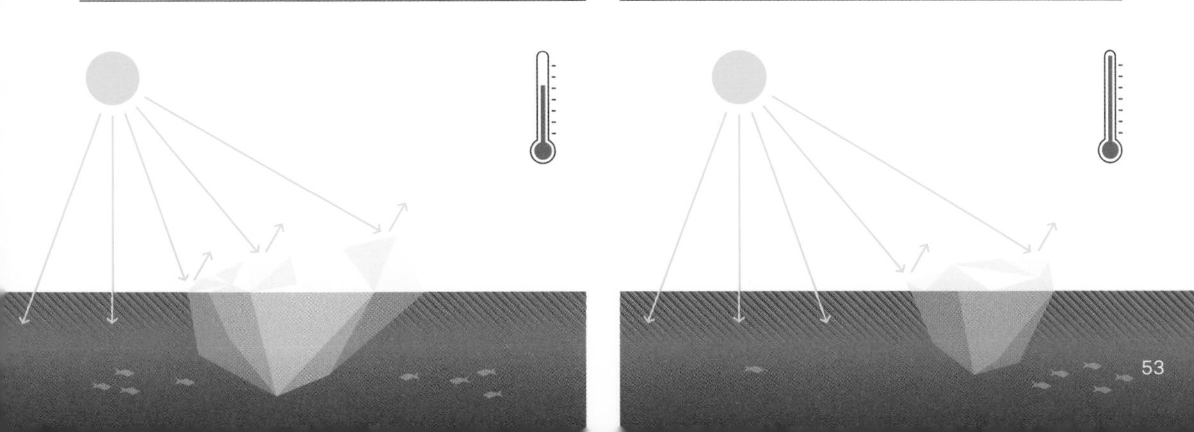

Less solar radiation is reflected back into space.

Feedback contributes to rising temperatures.

Land Ice

An ice cap is a generally flat, ice-covered area of land smaller than 19,300 square miles (50,000 km²). They are mainly found in polar and subpolar regions.[1] If the surface area is larger, it is referred to as an ice sheet or land ice.[2] At present, there are two ice sheets on Earth: the Antarctic and Greenland ice sheets.[3]

Muir Glacier, Alaska, 1941

One of the best-known effects of climate change is probably the shrinking mountain glaciers and ice caps. Higher temperatures and various local factors—such as annual snowfall rates—affect glacier retreat.[4,5] But mountain glaciers and ice caps only account for a fraction of the world's ice mass. The vast majority—more than 99 percent of the global land ice mass—is in the ice

This page: Field, William Osgood. 1941. Muir Glacier: From the Glacier Photograph Collection. Boulder, Colorado USA: National Snow and Ice Data Center. Digital media.

Opposite: Bruce Molnia, USGS. Public domain

Muir Glacier, Alaska, 2013

sheets of Greenland and Antarctica. Almost all glaciers observed worldwide are gradually losing mass.[6]

Not only are the glaciers retreating, there is also less and less snow coverage in the northern hemisphere. Since 1966, the surface area of land covered with snow has decreased at an average rate of 82 square miles (213 km^2) per year.[7]

Greenland's Ice

The Greenland ice sheet is the second largest on Earth after the Antarctic ice sheet.[1] It covers almost the entire land area of Greenland, and in many places it is more than 2 miles (3 km) thick.[2]

When ice breaks away from the edge of glaciers and plunges into the sea, it is referred to as "calving."[9]

①

Unlike the Arctic sea ice, the ice sheet of Greenland is grounded on land. The melting of Greenland's ice sheet is one of the reasons for rising sea levels. Once the entire mass of the ice sheet is gone, the global sea level will rise by over 23 feet (7 m).[1] Between 2002 and 2016, the melting of the Greenland ice sheet was responsible for a sea level rise of approximately 0.8 millimeters per year.[3,4] This corresponds to an average annual loss of about 280 gigatons (309 billion tons) of ice. This loss arises primarily from the increased calving of icebergs, where chunks of ice break away to form icebergs ❶,[5,6] and from ice melting at the surface of the sheet.[7] It should be noted that the Greenland ice sheet has been losing mass at an accelerating rate in recent years.[3,4,8]

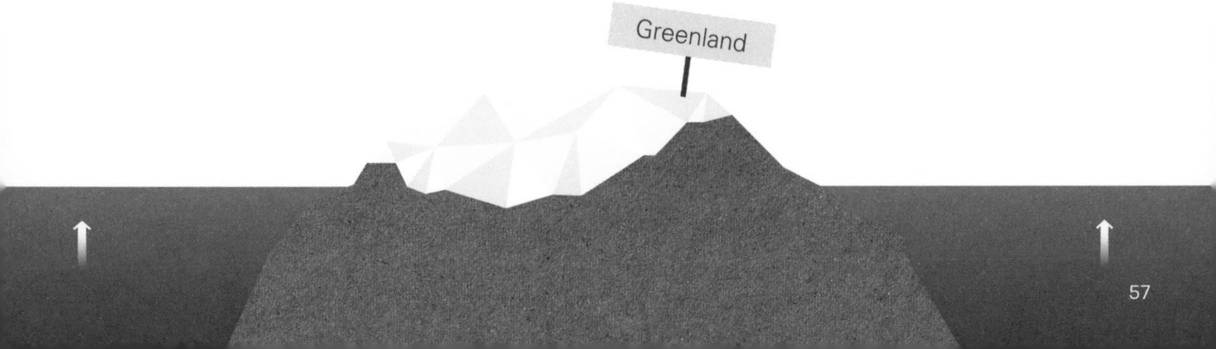

Greenland

Antarctica

Antarctica is covered by the largest sheet of ice on Earth.[1] Most of the ice in Antarctica is on land, and the rest lies as connected, floating coastal ice shelves.[2,3] There is so much ice in Antarctica that the sea level would rise by about 190 feet (58 m) if the entire ice sheet were to melt.[1] Unlike in the Arctic, the area covered by sea ice in the Antarctic region showed an annual average increase of 0.16 percent between 1979 and 2016.[4] However, the ice sheet is declining in terms of overall mass; while a slight increase in inland ice can be observed in some parts of East Antarctica due to increased snowfall,[5] mass is being lost in West Antarctica.[3] To a large extent, this loss is due to the melting of the West Antarctic ice shelf by the relatively warmer seawater beneath.[6] As the shelf gets thinner, it becomes less effective in holding up the flow of ice from the interior. This increases the speed of the ice streams' flow away from the land, meaning that they transport more ice seaward than is replenished by new snowfall.[7–10] In many parts of West Antarctica, the ice farther inland is grounded well below sea level, so when the ice retreats there will be a much larger area vulnerable to the warmer seawater. This might further accelerate the melting process and thus the speed at which the ice streams flow out to sea.[10–11]

Overall, from 2003 to 2016, there was an annual mass loss of roughly 141 gigatons (155 billion US tons).[7]

Opposite: Average yearly change in the overall mass of Antarctic sea ice between 2003 and 2016

Source: Sasgen et al. (2017)

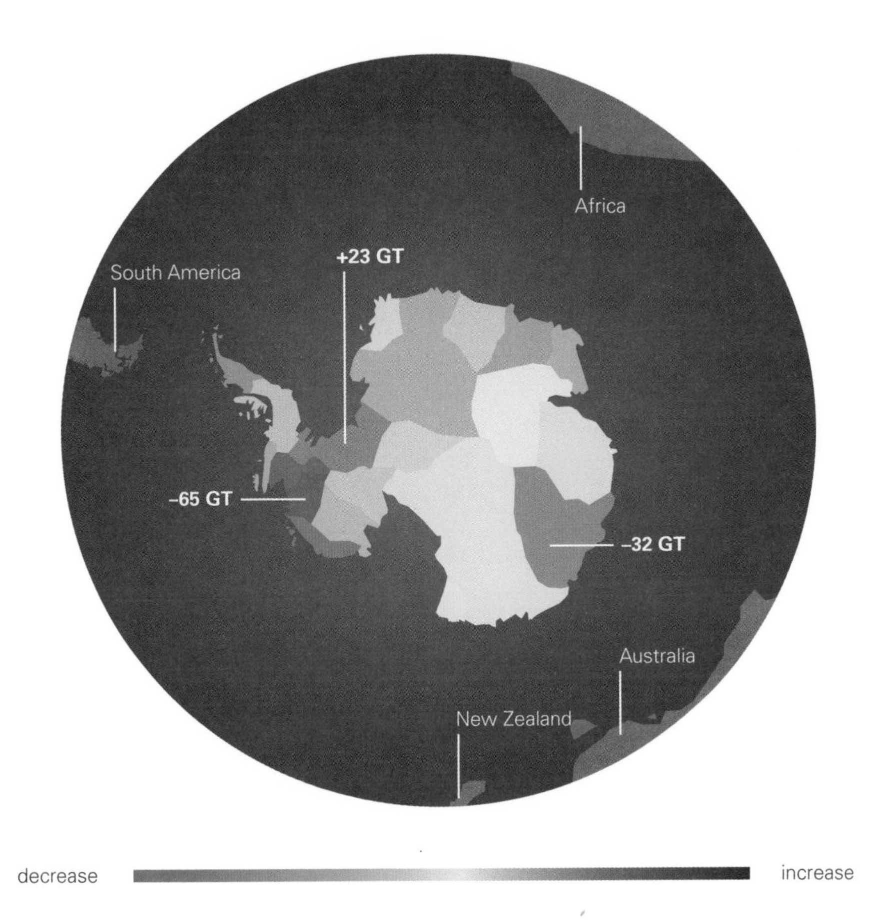
Africa
South America
+23 GT
–65 GT
–32 GT
Australia
New Zealand
decrease
increase

Melting Ice and Rising Sea Levels

As the term implies, "land ice" is ice located on land. When this ice melts, the meltwater flows into the sea, thereby causing the sea level to rise (p. 72). If all of the world's land ice were to melt, the sea level would rise by approximately 217 feet (66 m).[1] Sea ice and ice shelves are a different matter, because they are both already in the water and, if the water and the

When sea ice melts, it produces virtually the same quantity of water as it had previously displaced.

ice have an identical salt content, the melting sea ice will produce the same amount of water as was previously displaced by it (as is shown in the illustration below). However, even though sea ice and ice shelves have different salt content from seawater, their ice displaces only slightly less water than it would when melted. For this reason, if all the sea ice and the ice shelves melted, the sea level would rise by only about 1.6 inches (4 cm). The ice shelves would contribute most to this effect, about 1.4 inches (3.6 cm).[2]

The melting of Arctic sea ice therefore has little or no effect on the average sea level.

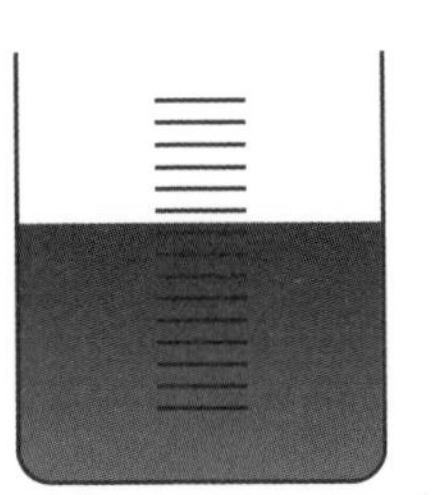

Permafrost

Permafrost is defined as ground that remains at a temperature of 0°C (32°F) or lower for least two consecutive years.[1] *Permafrost is found in cold parts of the world such as Siberia, Canada, Alaska,*[2] *or in mountainous regions,*[3] *and covers approximately 24 percent of the land surface area of the northern hemisphere.*[4]

As a result of global warming, permafrost is beginning to thaw in the polar summer, and this thaw is lasting longer and spreading deeper.[5] Preserved within the permafrost soil are thousands of years of plant remains and animals. If permafrost thaws they are exposed to microbial decomposition processes that convert the carbon stored in the plants and animals into carbon dioxide (CO_2) and methane (CH_4). Both gases can then seep out into the atmosphere ❶.[6] But, higher temperatures also lead to increased plant growth. In the short term, plants can absorb more CO_2 than is released by the thawed soil ❷—but not in the long term ❸.[7] This further accelerates global warming, which will probably cause the permafrost to thaw even more. This self-reinforcing process is known as the permafrost carbon feedback (PCF). It will probably cause Earth to heat up more quickly than would be expected from human emissions alone.[6]

When permafrost thaws, it releases greenhouse gases that accelerate global warming.[6]

CH_4
CO_2
CO_2
①
CH_4
CO_2
CO_2
②
CH_4
CO_2
③
CO_2
CH_4
CO_2
CO_2

Other Consequences of Thawing Permafrost

Permafrost is structurally stable thanks to its frozen water content. When it thaws, the subsoil becomes unstable, which can cause damage to infrastructure such as buildings, pipelines, or the transport network.[1,2] Destabilized ground can also trigger landslides.[3] Permafrost in mountainous regions (like the Alps) has a stabilizing effect on rock faces; here, thawing increases the risk of rockfall.[4]

The combination of thawing permafrost and diminishing sea ice and the increase in air and sea temperatures might lead to increased coastal erosion.[1,5] This process is happening now, averaging a loss of 1.6 feet (0.5 m) per year, although there is a lot of variation.[6] For example, on some parts of the coast of Alaska, the land is being eroded at an average rate of 44.3 feet (13.5 m) per year.[7]

CHAPTER 4

THE OCEANS

The oceans cover more than 70 percent of Earth's surface. They are of enormous importance for maintaining our climate, as they transport vast quantities of heat around the world.[1,2] They also act as a buffer against global warming, because they absorb some human-made CO_2 emissions as well as a large part of the energy that is retained on Earth as a result of the human-made greenhouse effect.[3,4]

Impact on the Oceans

From 1971 to 2010, the oceans absorbed 93 percent of the energy that had been kept within Earth's atmosphere as a result of human-made climate change.[1] As a result, the average surface temperature of the oceans has risen—by 0.8°C (1.4°F) in the period from 1880 to 2015—and the deeper layers beneath have also warmed.[1,2] The oceans are becoming more acidic as they absorb about 22 percent of human-made CO_2 emissions.[3] In this way, the oceans are attenuating global warming, but at the expense of their sea life, which is being increasingly affected by the warming and acidification of its habitat (p. 100).[4]

As gases are less soluble in warm liquids than in cold ones, the oceans are becoming less able to absorb human-made CO_2 emissions.[5] Consequently, they are becoming a less effective buffer to global warming.[6] The oxygen level in the oceans is also decreasing, placing an additional strain on sea life.[7]

Where does the excess energy from global warming go?[1]

93% oceans

3% continents

3% melting ice (Arctic sea ice, ice sheets, and glaciers)

1% atmosphere

22%

of human-made CO_2 emissions absorbed by the ocean[3]

Oceans are becoming more acidic.

CO_2

CO_2

CO_2

CO_2

CO_2

CO_2

CO_2

CO_3^{2-}

H_3O^+

HCO_3^-

H_3O^+

Water Vapor Feedback

Starting point

Warming causes increased water to evaporate.

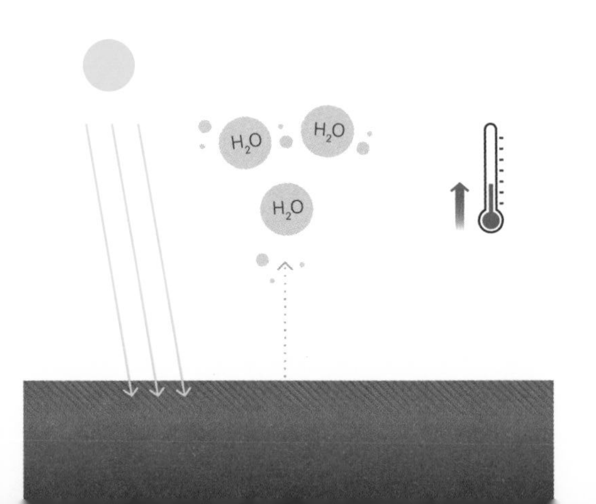

Warm air can absorb more water vapor than cold air. So when the air temperature rises, the water vapor content in the atmosphere increases too.[1] Since water vapor also functions as a greenhouse gas, the additional water vapor in the atmosphere intensifies the greenhouse effect and adds to the ever-rapid rise in temperature.[2] Thus a self-accelerating process known as water vapor feedback is established, which speeds up global warming.[3]

The greenhouse effect is intensified.

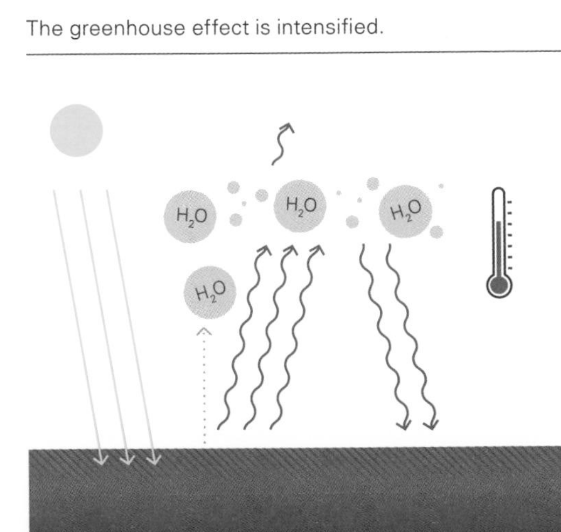

Feedback increases warming.

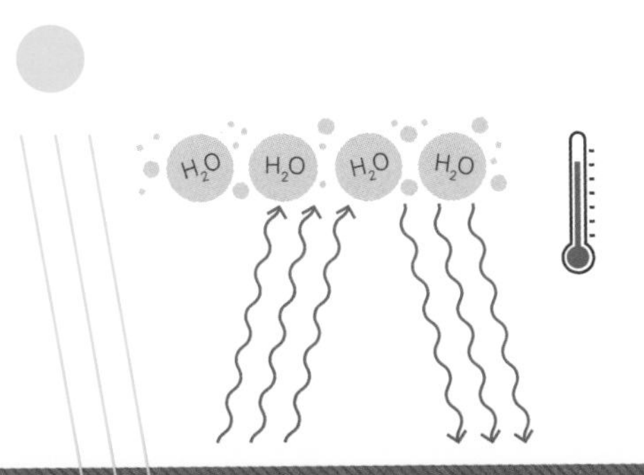

Rising Sea Levels

Because the oceans have warmed dramatically as a result of climate change, seawater is expanding in volume, thus causing sea levels to rise.[1,2] Glaciers and ice sheets are melting with the rising air temperature, and, as a result, global sea levels rose by a total of 9 inches (23 cm)

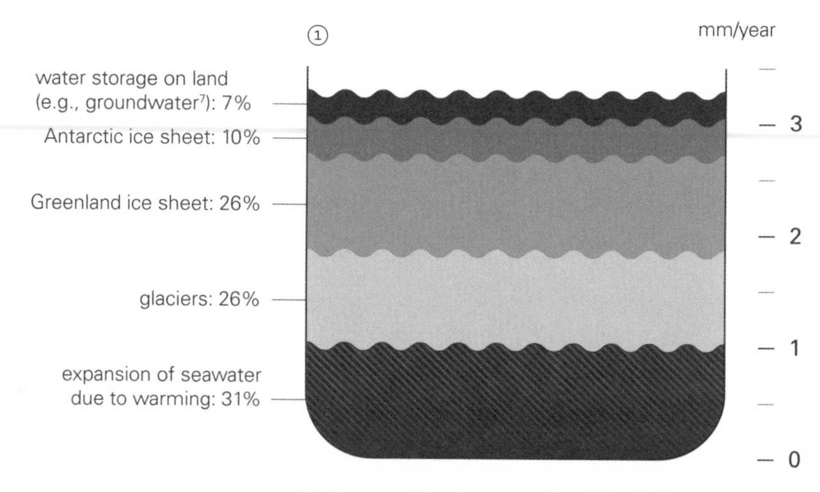

Factors contributing to sea level rise in 2014[4]

between 1880 and 2013.[3] In 2014 the rise was approximately 3.3 mm ❶.[4] Greater sea level rises than we are seeing today have occurred in the past, when large ice masses melted during the transition from ice ages to warmer periods (at a rate of 10–15 mm per year).[5] But if we look at the average sea level rise over the last fifteen years, compared to the past century and the past 2,000 years, we see that sea levels are rising more and more rapidly ❷.[5,6]

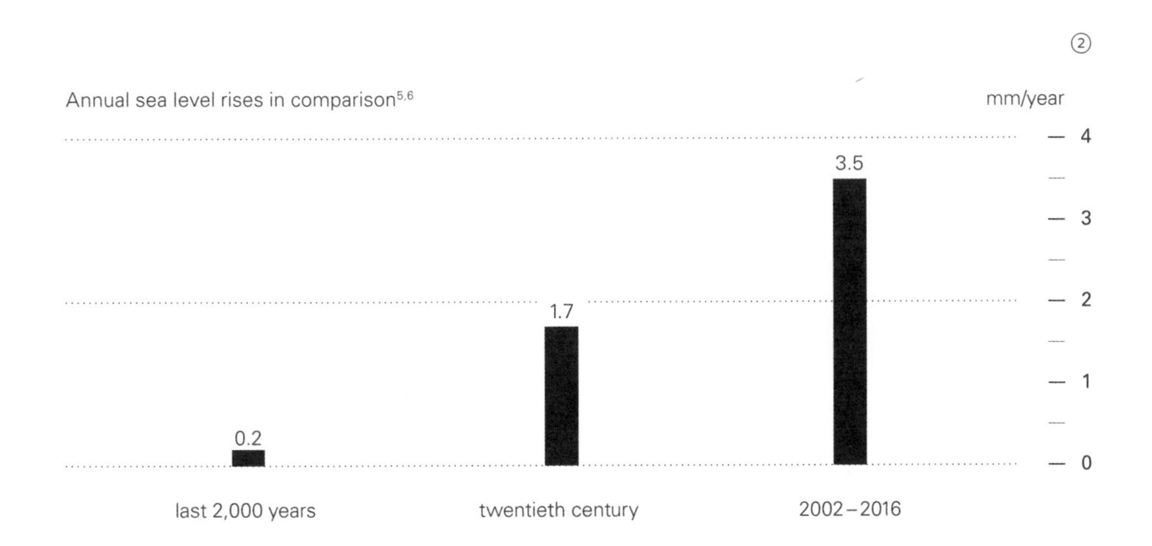

Changes in Oceanic Circulation

The Atlantic Meridional Overturning Circulation (AMOC) is a part of the "global conveyor belt" (p. 16).[1] Together with the Gulf Stream ❶ and the North Atlantic Current ❷, it transports large quantities of heat from the tropics to the North Atlantic, thereby helping to maintain the moderate climate in northwest Europe.[2] The warm, salt-rich surface water flowing north releases its heat into the atmosphere in the North Atlantic. This leaves the seawater cooler and denser, causing it to sink and then flow back southward in the deep ocean.[3] In the south, different water masses combine in the deep ocean, which, helped by surface winds, brings the deeper water back to the surface.[4] The AMOC cycle can then begin again.

As climate change progresses, Greenland's ice sheet continues to melt (p. 56). Because the meltwater is low in salt content, it reduces the density of the surface water in the North Atlantic. This could mean that the water masses from the south will no longer sink deep enough to be transported back, potentially weakening the AMOC cycle.[5] So far, there is no clear indication that this is happening, but model simulations show that the AMOC could weaken by between 11 percent and 34 percent by the end of the twenty-first century, due to the increase in human-made greenhouse gas emissions.[6,7,8] This may have consequences on both sides of the Atlantic, from rising sea levels and more hurricanes on the East Coast of the US[9,10] to a change in wind circulation patterns and, consequently, more severe storms in Europe.[11]

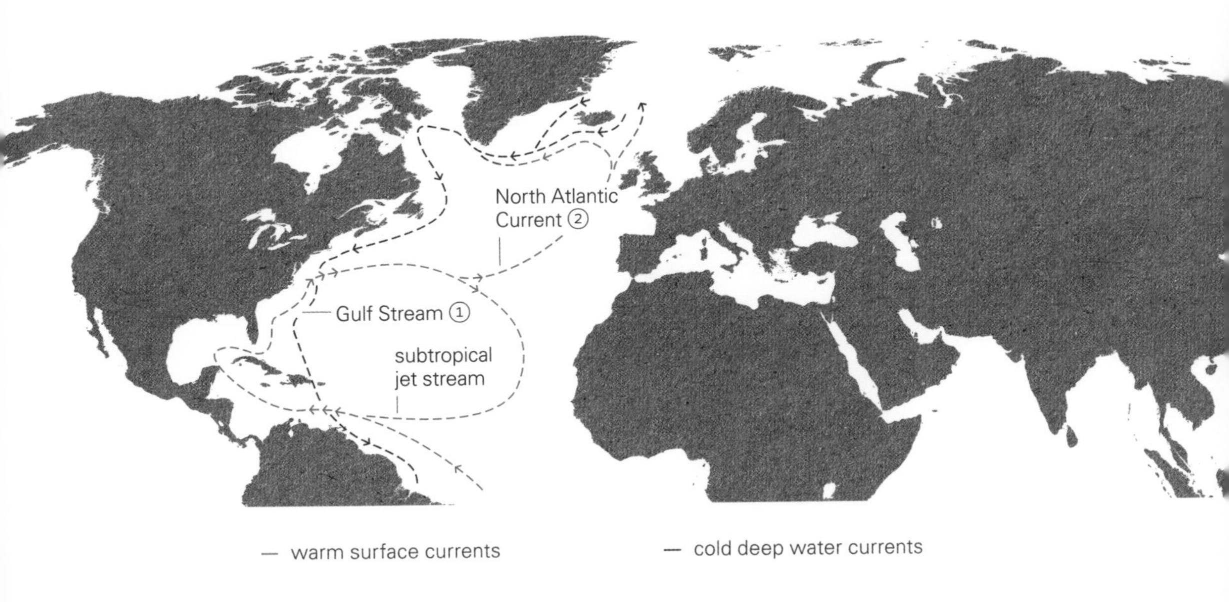

— warm surface currents — cold deep water currents

CHAPTER 5

WEATHER AND CLIMATE EXTREMES

Weather and climatic events are usually defined as extreme if they exceed certain thresholds or there is a very low probability of their occurrence. However, there is no universal definition for extreme weather or climate conditions.[1]

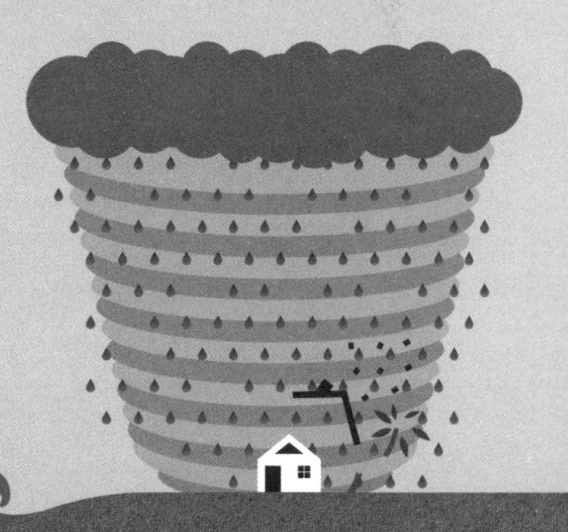

Extreme Hot and Cold Weather

Climate change is increasingly causing record-breaking high temperatures[1] and heatwaves.[2] Between 1951 and 1980, less than 1 percent of Earth's land surface experienced unusually high summer temperatures. "Unusually high" temperatures in this context are defined as having a "maximum theoretical probability of 0.13 percent of occurring," which is to say, a very low probability. These hot-weather events, once very rare, occurred on approximately 10 percent of Earth's land surface in the period from 2001 to 2010.[3] In addition, the average duration of the wildfire season increased by around 19 percent between 1979 and 2013 worldwide.[4] It should be noted, though, that wildfires are often trigged by humans, either through carelessness or arson.[5]

In contrast to heatwaves, extremely cold spells are becoming both less frequent and less severe.[6] This is due to the shift in temperature distribution resulting from climate change, as can be seen in the illustration opposite. In spite of global warming, extremely cold periods can still occur at a local level—although they may be expected to occur less often and in a weaker form.[7]

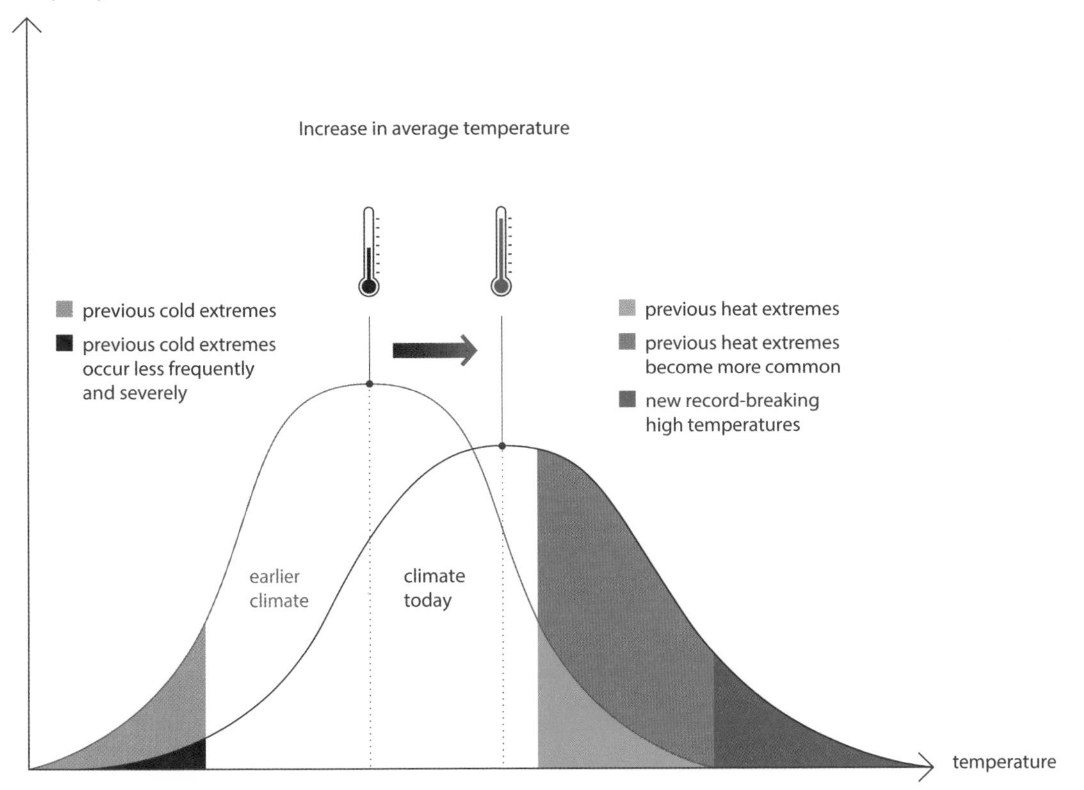
frequency
Source: Hansen & Sato (2016)
Increase in average temperature
previous cold extremes
previous cold extremes occur less frequently and severely
previous heat extremes
previous heat extremes become more common
new record-breaking high temperatures
earlier climate
climate today
temperature

Rainfall and Flooding

Because higher temperatures allow the air to hold more water vapor, the water vapor content in Earth's atmosphere is increasing. Higher temperatures also allow more water to evaporate, given sufficient humidity (for example, over the oceans). This intensifies the water cycle, meaning that higher precipitation levels may be expected.[1,2] Water vapor does not usually fall as rain in the same place that it evaporated,[1] and this, combined with shifting atmospheric circulation patterns, results in an increasingly uneven distribution of rainfall.[3] Dry regions such as the subtropics are in many cases becoming even drier, while many humid regions like the midlatitudes and tropics are becoming even wetter.[3,4,5]

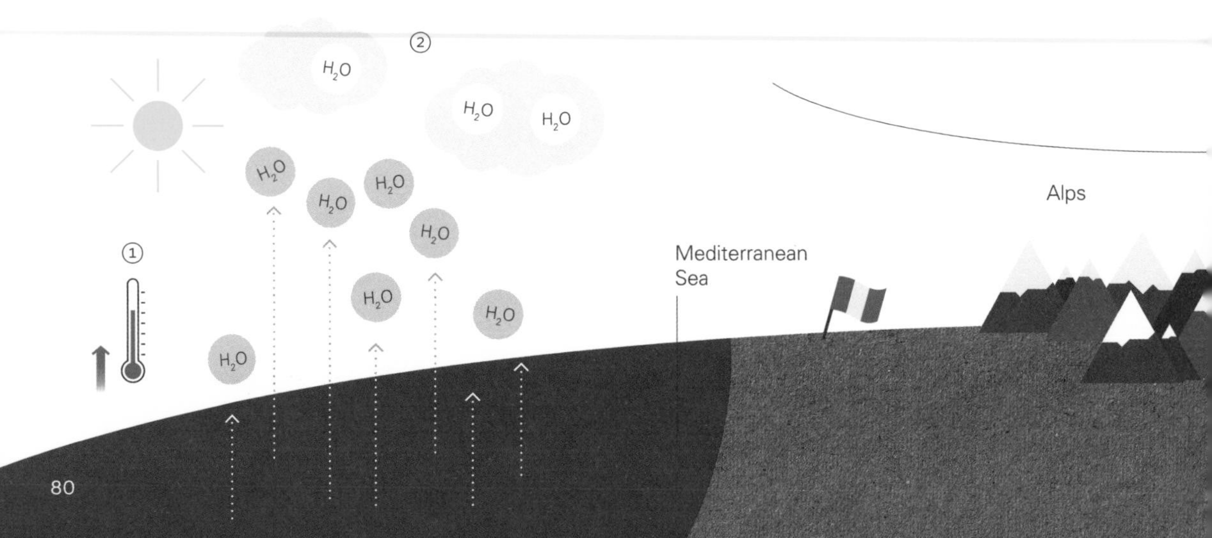

On average, intense rainfall events are likely to become more frequent around the world,[2] and more severe in both dry and humid regions.[6] At present, about 18 percent of the world's heavy rainfall events on land can be attributed to global warming.[7] That said, the rising frequency of these extreme precipitation events varies considerably from place to place.[2] For example, the Mediterranean Sea has warmed significantly in recent decades ❶, so more water is evaporating over it ❷; if high and low pressure areas come together in a particular configuration (known as Vb track) ❸, this water vapor will be transported northward, which may lead to an increased frequency of intense precipitation events and flooding in Central Europe ❹.[8]

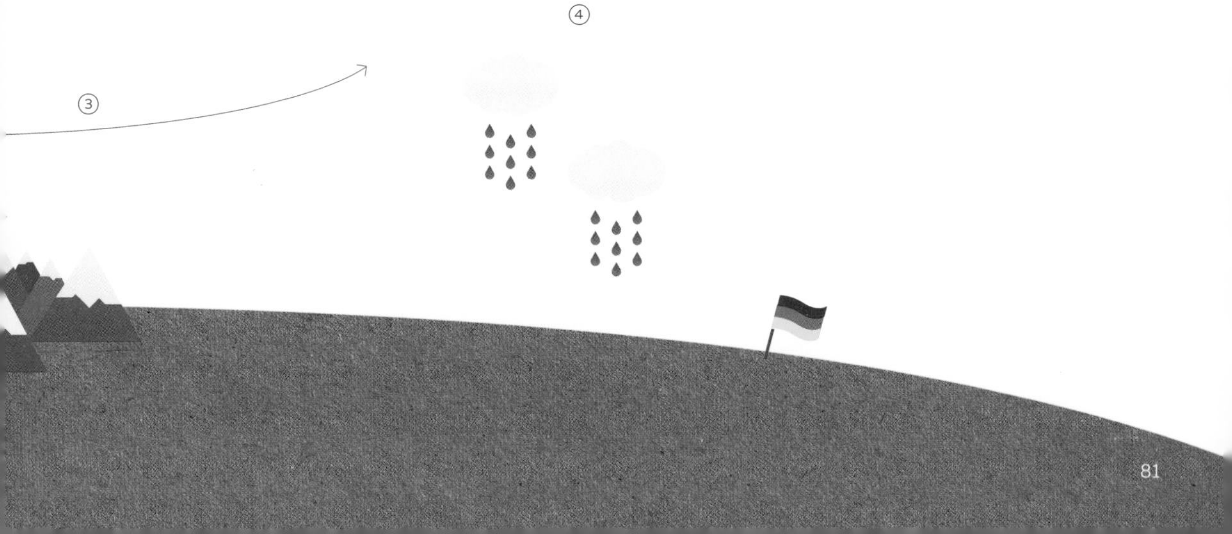

④

③

Droughts

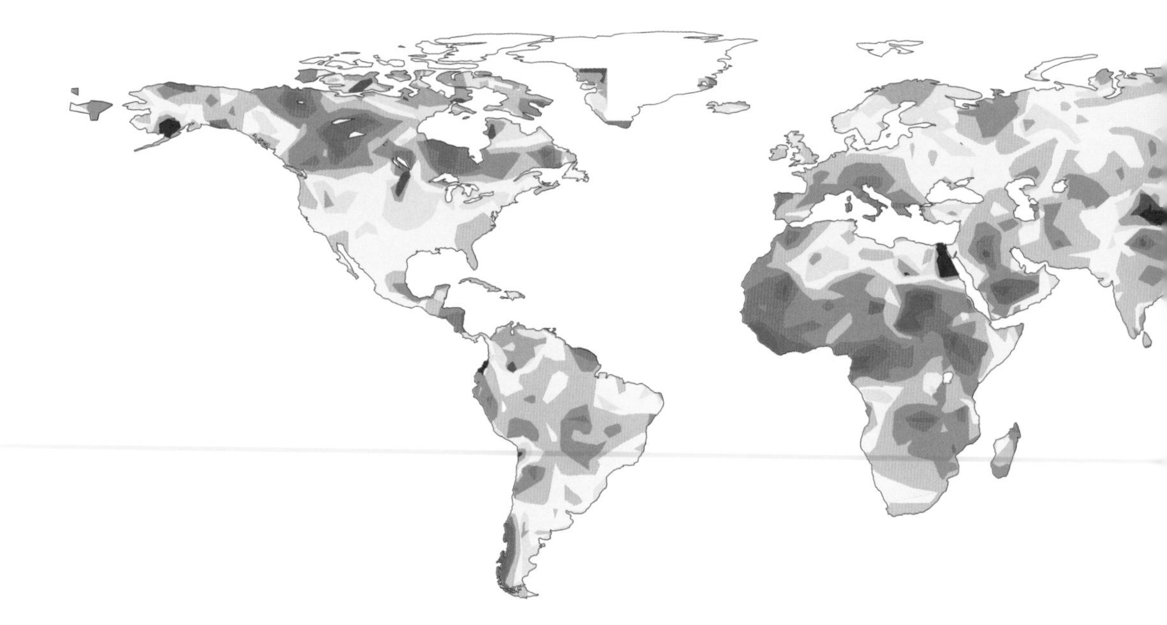

Regions where aridity increased or declined between 1950 and 2012

Adapted with permission from Aiguo Dai and Tianbao Zhao, "Uncertainties in historical changes and future projections of drought. Part 1: estimates of historical drought changes," Clim. Change *144 (2017): 519–33. Copyright © Springer Nature*

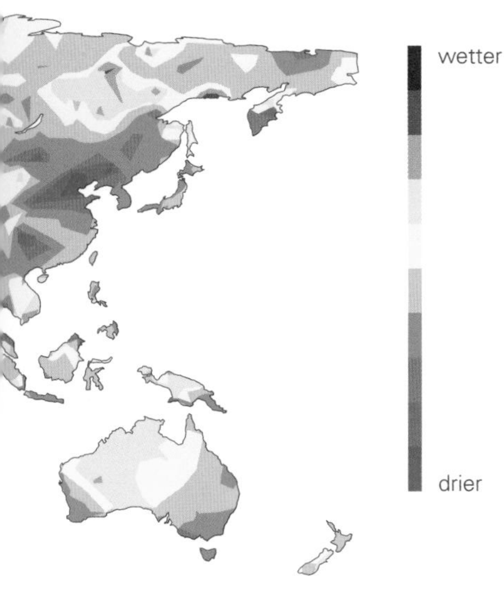

Since the concept of drought has a number of different scientific definitions, it is difficult to make a generalized statement about global drought trends. But we do know that natural factors are crucial in determining where and when droughts occur, and global warming increases the localized evaporation of moisture from the soil, thus increasing the probability of droughts occurring both more quickly and more severely.[1] Additionally, since the middle of the twentieth century, the total area of arid land around the world has grown as a result of climate change. This is particularly the case in Africa, southern Europe, East and Southeast Asia, and in many parts of the northern, middle, and high latitudes. This could increase the risk of droughts in the future.[2,3]

Tropical Cyclones

As things currently stand, we can only demonstrate a direct correlation between the increasing frequency of tropical cyclones and climate change for the most severe of these storms.[1,2] In general, it is estimated that climate change will lead to fewer tropical cyclones overall. Weak storms will probably decrease in number, while more severe storms will become increasingly frequent.[1,3]

Tropical cyclones form over the ocean when the water temperature rises above 26°C (79°F), as they are powered by warm, humid air (latent heat).[4,5,6] Climate change is causing the temperature of the sea surface to rise ❶ and consequently evaporation is increasing in turn, thus feeding more energy to the storms ❷, and increasing the chance of more severe storms occurring ❸.[4,7] It is also to be expected that future cyclones will be accompanied by heavier rainfall, due to the increased airborne water vapor.[3]

On the other hand, global warming is allowing for less upward and downward movements of air (greater atmospheric stability) and strengthening instead stacked horizontal air currents whose winds vary in both direction and speed. This is the case, for example, at higher altitudes over regions of West and Central Africa, where Atlantic hurricanes form. This stability can inhibit the formation of tropical cyclones, so their overall number could decline.[8,9]

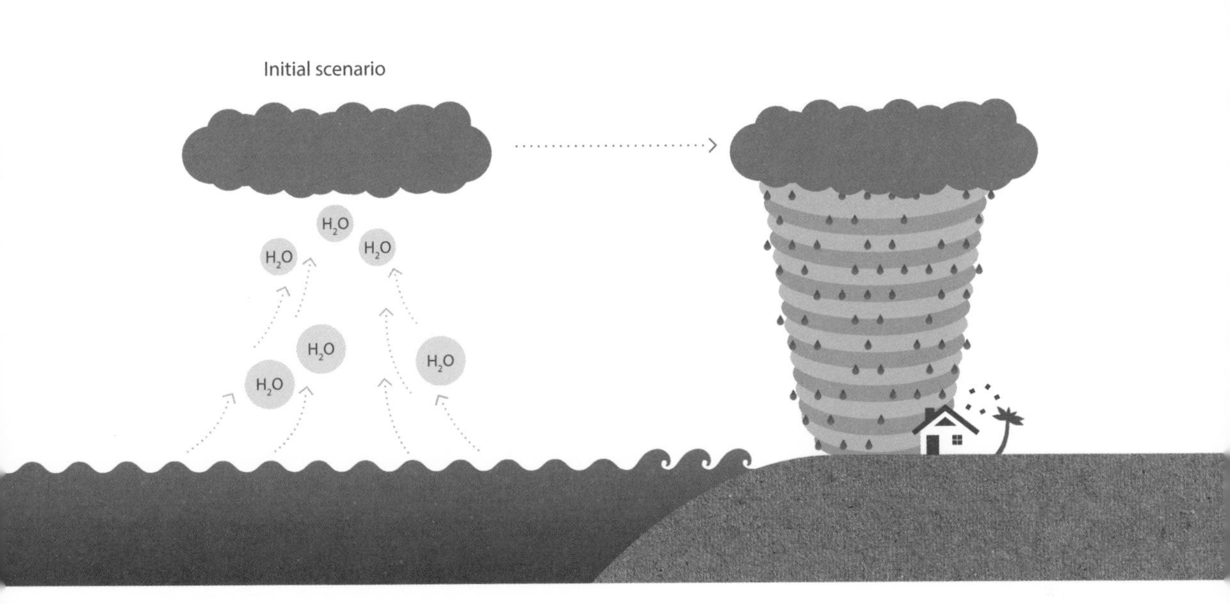

Intensification as a result of climate change

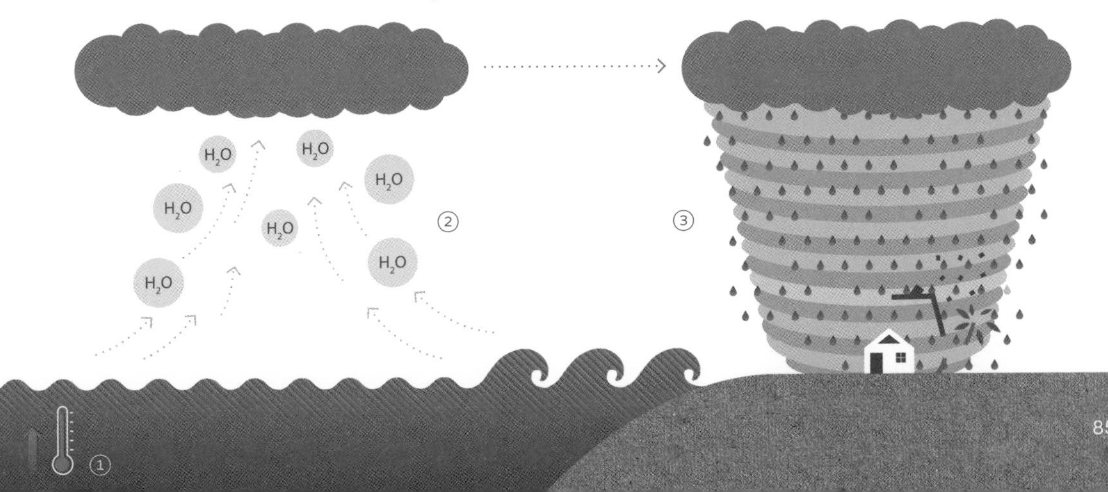

Thunderstorms

Thunderstorms often come accompanied by extreme weather phenomena such as heavy rainfall, hail, strong winds, or tornadoes. They can cause billions of dollars' worth of damage.[1,2,3]

Thunderstorms are formed by a combination of factors. All thunderstorms require three conditions before they can develop: sufficient moisture, unstable air masses, and a "lifting mechanism" triggering the air to rise.[4] Severe thunderstorms additionally require substantial changes among stacked horizontal air currents, whose wind direction and/or speed varies with altitude ("vertical wind shear").[5] In particular, rising, warm, and humid

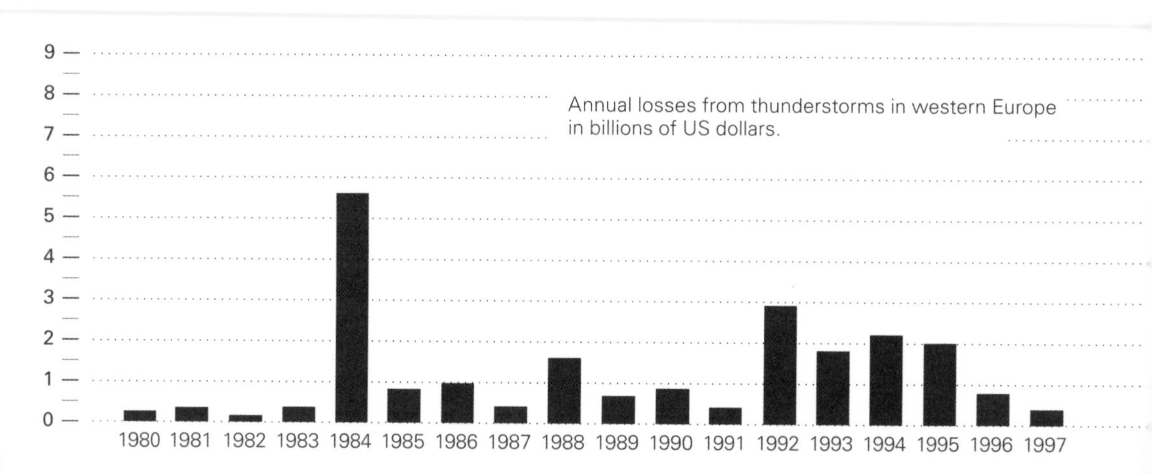

Annual losses from thunderstorms in western Europe in billions of US dollars.

air plays a vital role in supplying the energy needed for thunderstorms to form. The water vapor it contains condenses at high altitudes, thereby releasing thermal energy and thus boosting the thunderstorm formation process.[6] As the air gets warmer with climate change, and can therefore hold more moisture, more energy can be released when it condenses. We can expect an increase in the number and intensity of thunderstorms.[7,8] Given the variety of factors influencing thunderstorm formation and insufficient data on thunderstorms worldwide, it is difficult to generalize about global trends.[9] It is easier to forecast locally.

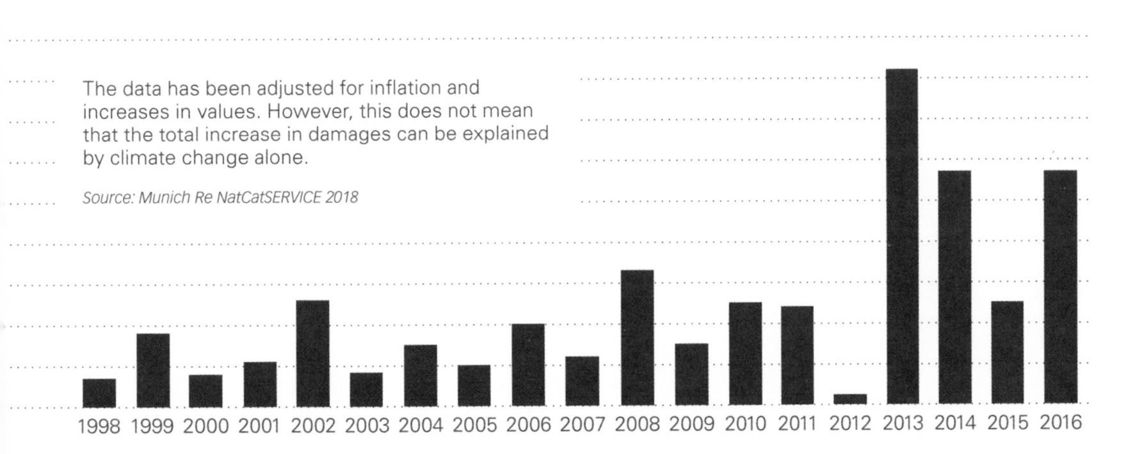

The data has been adjusted for inflation and increases in values. However, this does not mean that the total increase in damages can be explained by climate change alone.

Source: Munich Re NatCatSERVICE 2018

CHAPTER 6

ECOSYSTEMS

Interdependent organisms and the habitat or physical environment in which they are found form a biological community based on their interactions. This is referred to as an "ecosystem."[1]

The term "phenology" refers to observable stages in plant and animal development that recur periodically over the course of a year.[2]

"Biodiversity" refers to the variety of life forms and ecosystems, as well as interactions between individuals and ecosystems, and genetic diversity within species.[3,4,5]

Seasons, Vegetation Zones, and Climate Zones

The term "phenology" refers to observable stages in plant and animal development that recur periodically over the course of a year[1]—for example, the breeding season for birds or the flowering of plants.[2]

Higher temperatures result in phenological changes,[3] such as birds breeding earlier in the year, or plants blooming earlier in the year.[5] Over the last few decades, the phenological spring in the northern hemisphere has started on average approximately 2.8 days earlier each decade.[6] These seasonal shifts increase the closer one comes to the poles.[3] Climate change is also causing a shift in vegetation zones.[7] The tree line in the northern hemisphere, for instance, is moving northward[8] and reaching higher into the mountains.[9] The retreat of Arctic and Alpine ecosystems also illustrates this shift in vegetation zones.[10]

Climate zones have changed in much the same way. Between 1950 and 2010, roughly 5.7 percent of Earth's land surface shifted toward warmer and drier climate zones.[11] Climate change also produces new combinations of climate elements which are as yet unknown, making it very difficult to estimate their consequences.[12]

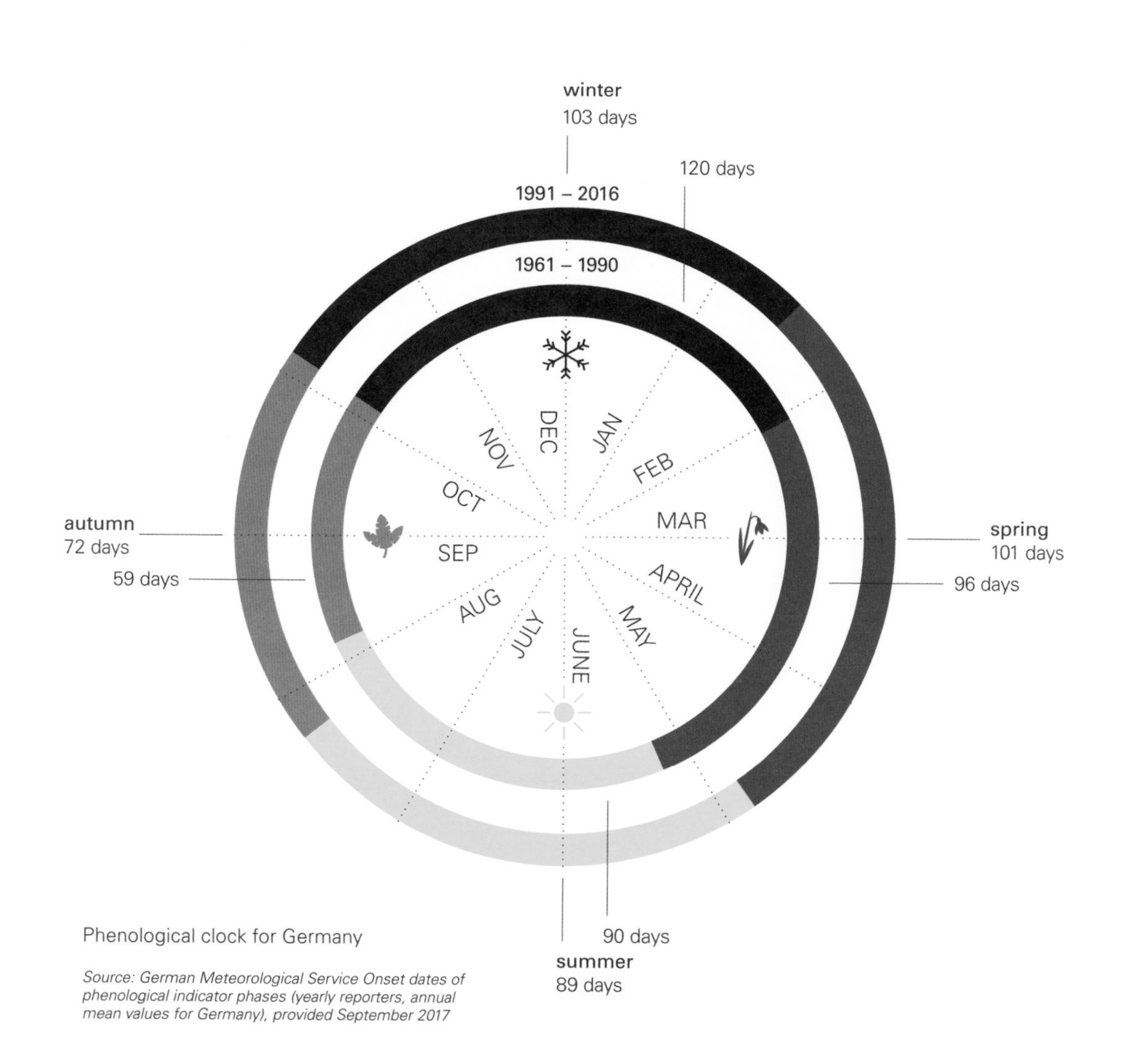

Phenological clock for Germany

Source: German Meteorological Service Onset dates of phenological indicator phases (yearly reporters, annual mean values for Germany), provided September 2017

Animals and Plants

Generally speaking, animals and plants are well adapted to the climatic conditions of their habitat. If the climate changes, there are attendant changes in the composition of species communities—and as a consequence there are often changes to the whole ecosystem.[1]

In principle, there are three ways in which a species might respond to a changing climate:

1. They might adapt to the changes,[2] coping very well and even increasing their numbers, as the bark beetle has done in many parts of Europe.[3–6]

2. They might move their habitat, as species like certain kinds of butterfly do. Usually this migration occurs in the direction of the poles or to higher elevations—to escape temperatures that are too high. Land animals and plants have been found to be moving to altitudes roughly 36 feet (11 m) higher each decade, and traveling by around 10.5 miles (17 km) farther toward the poles.[7]

3. They might be unable to adapt to the changes, which can reduce their range and may ultimately result in their extinction.[8] The faster the changes take place, the greater the danger that animals and plants will not be able to adapt quickly enough to survive.[9]

If one species changes, there can be repercussions for the entire ecosystem. Examples include shifts in predator–prey or competitor relationships.[1]

Changes in the elevation profile of plants, butterflies, and birds due to climate change in Switzerland from 2003 to 2010[10]

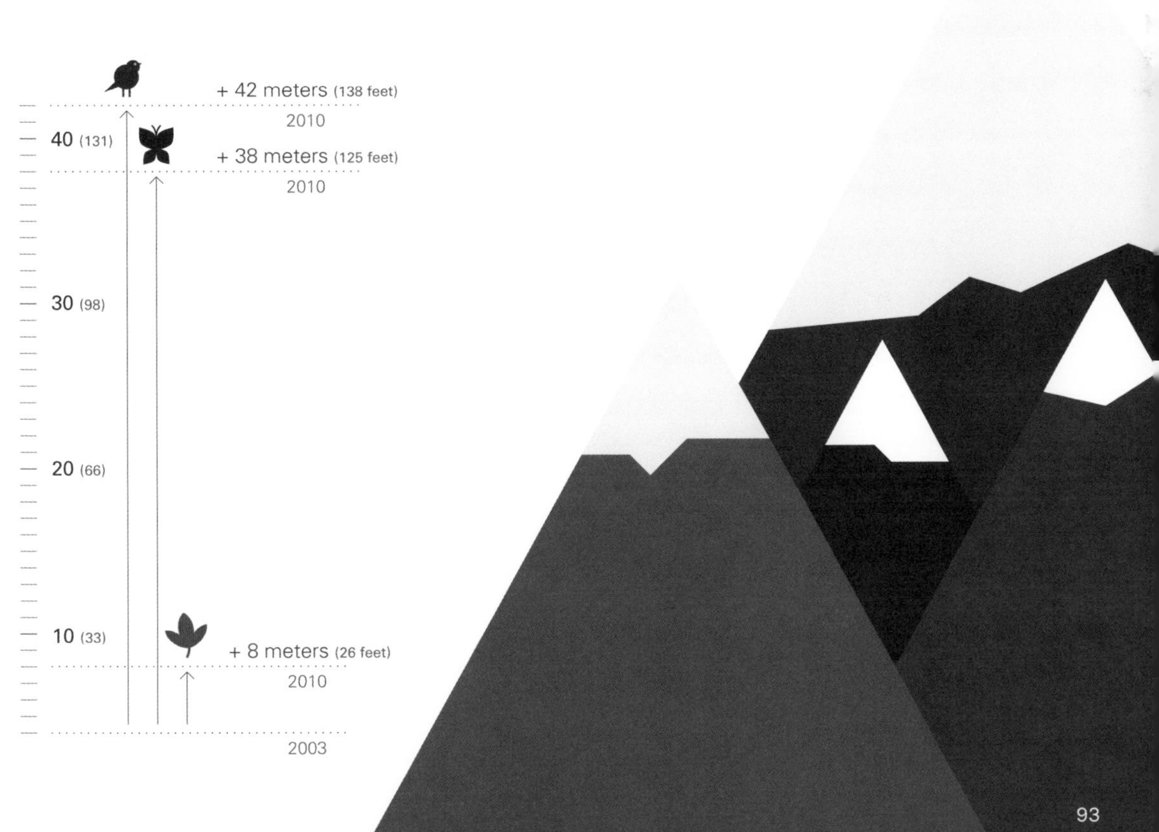

Biodiversity and Ecosystem Services

"Biodiversity" refers to the variety of life forms and ecosystems, as well as interactions between individuals and ecosystems, and genetic diversity within single species.[1,2,3]

High biodiversity improves the robustness of an ecosystem and, therefore, its ability to adapt to external events like climatic changes[4] or to diseases such as fungal attack.[5] High plant diversity also increases the productivity of an ecosystem.[6,7] This means that, assuming otherwise unchanged conditions, low plant diversity

Categories of ecosystem services

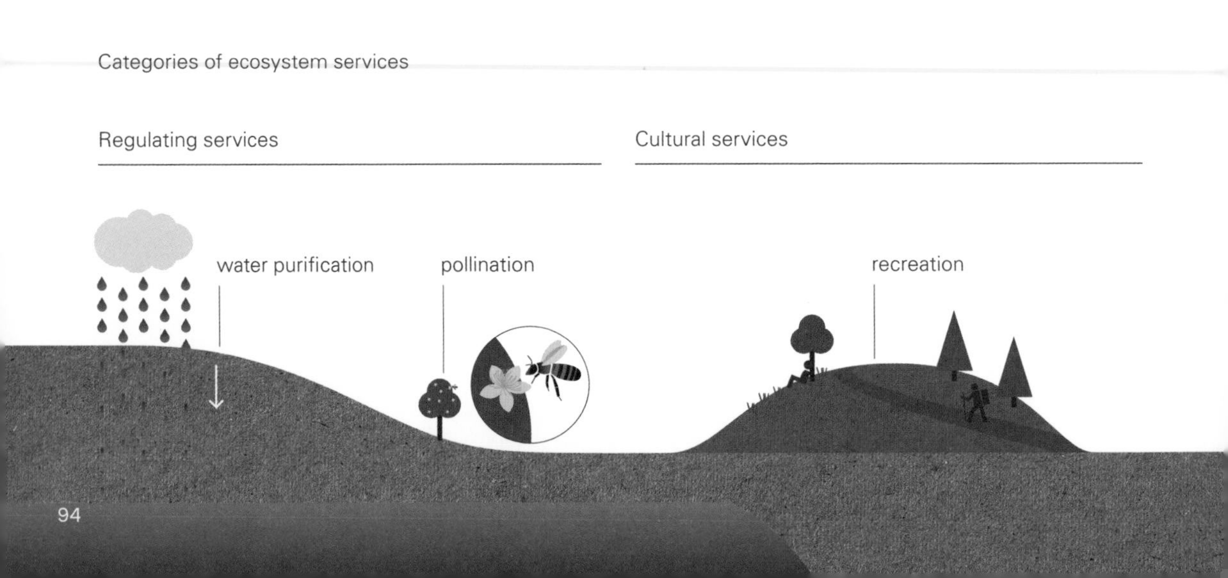

produces less biomass than a system with a high plant diversity.[6]

Overall, worldwide biodiversity is likely to decline because of global warming,[8] which could have a negative impact on "ecosystem services."[9] This term refers to all the services and properties of an ecosystem that benefit humans.[10] These services, which are essential for humans and generally cost nothing, can also be negatively affected by changes to species combination within an ecosystem.[11]

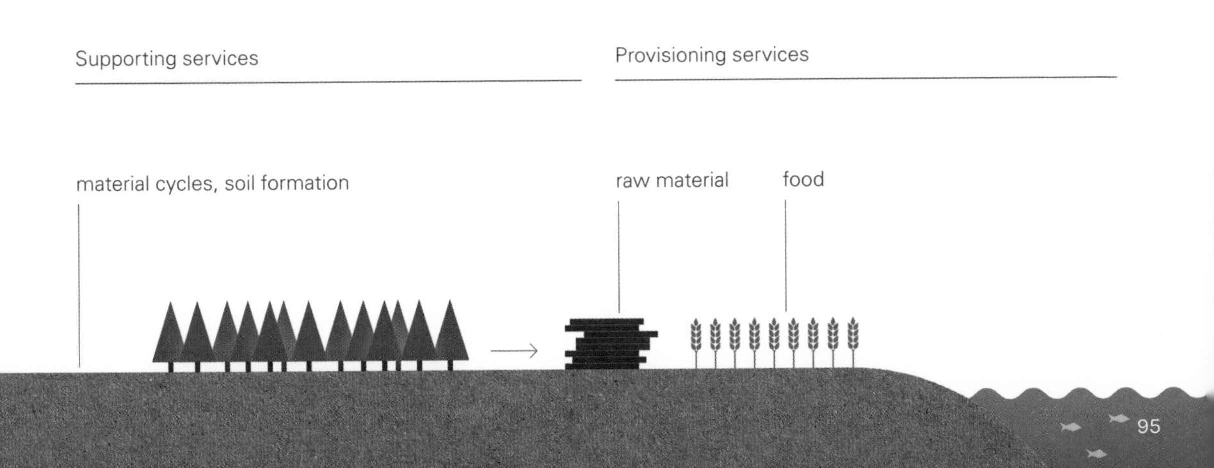

Pikas and Hummingbirds

Climate change poses a particular threat to animals that live in high altitudes or the northern latitudes. Such species have limited scope to escape climate warming and changes in their habitat.[1]

One interesting example is the American pika, a close relative of rabbits and hares, which can be found primarily in mountainous terrain in western North America.[2,3,4] They have been affected in a number of ways by climate change.[5] Pikas don't hibernate, but instead spend the winter in their burrows, where they feed on plants collected during the summer.[4] Snow cover insulates against temperatures that would otherwise be too cold for them. If the snow cover retreats, it can become too cold in their burrows, which can lead to the pikas freezing to death. The increasingly warm summers also pose a problem for these highly temperature-sensitive creatures.[2]
In response, the lower boundary of their habitat has been moving to increasingly higher, and therefore colder, elevations.[2,5,6] Shifts of this kind are particularly problematic if a species lives

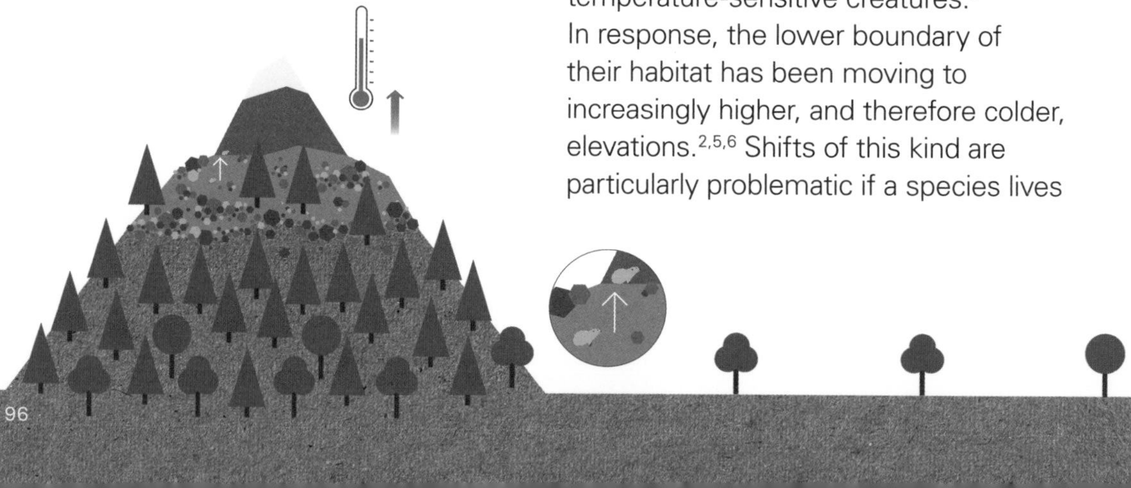

at the top of a mountain, and is thus unable to escape to higher altitudes.[5,7,8]

Phenological shifts can lead to changes in the interactions of plants and animals.[1] For example, there has been a shift in the phenological phases of the broad-tailed hummingbird and the plants that provide it with nectar. This hummingbird migrates from Central America northward to the mountains in order to breed in the summer. According to one study, between 1975 and 2011 the birds arrived at the northernmost breeding sites about 1.5 days earlier per decade, because of higher temperatures. Over the same period, the time at which most of the two observed plants began flowering shifted forward by roughly 2.8 days per decade: almost twice as fast. So the time between the birds' arrival and the flowering of their feeding source is reduced. This presumably also shortens the time available to the birds to build their nests and raise their young. If this time window continues to contract, it could seriously jeopardize the birds' chances of successful reproduction.[2]

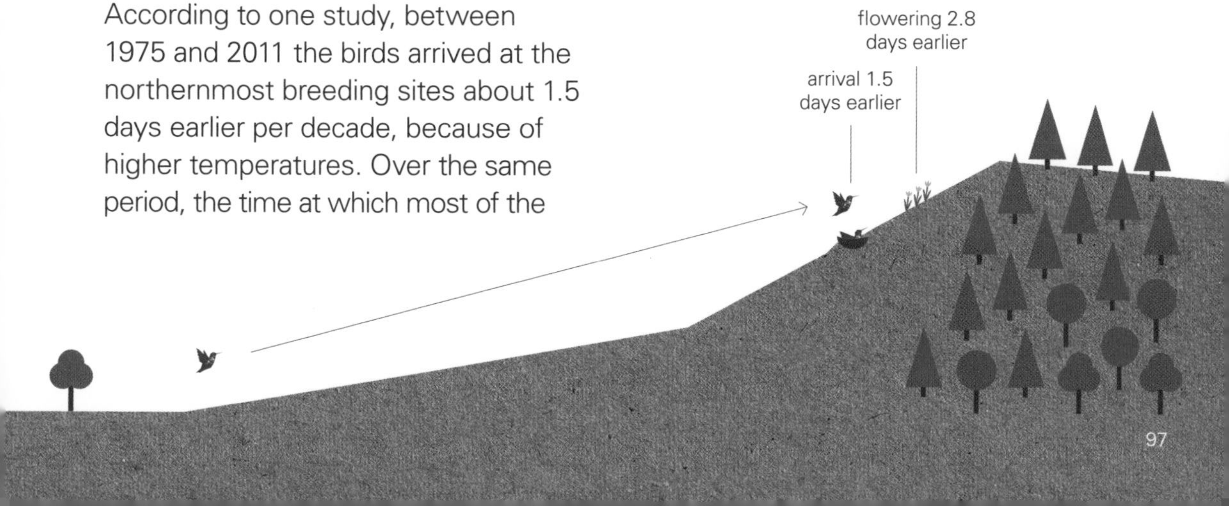

flowering 2.8 days earlier

arrival 1.5 days earlier

Polar Bears

The polar bear is the best-known Arctic creature affected by climate change. There are currently an estimated 25,000 polar bears in the Arctic and subarctic regions.[1,2]

The increasing speed at which sea ice is melting in the Arctic summer is shortening the time polar bears can use it as a platform to hunt seals. This reduced access to food can have a negative impact on the number of cubs they can successfully birth and rear. In extreme cases, the survival of adult polar bears is directly threatened.[3,4] While some polar bears have been observed eating berries and birds' eggs in their most southerly habitats in Canada, such alternative food sources are heavily dependent on local availability, and are not sufficient to meet polar bears' nutritional requirements.[4,5]

Projected population trends depend greatly on assumptions about future warming and associated changes in sea ice. For this reason, there is a great deal of uncertainty about how severe the impact will be on individual polar bear populations and over what time period. However, one thing is clear: If the sea ice retreats, the number of polar bears will decline.[4,6,7,8]

Source: IUCN / Polar Bear Specialist Group (2017)

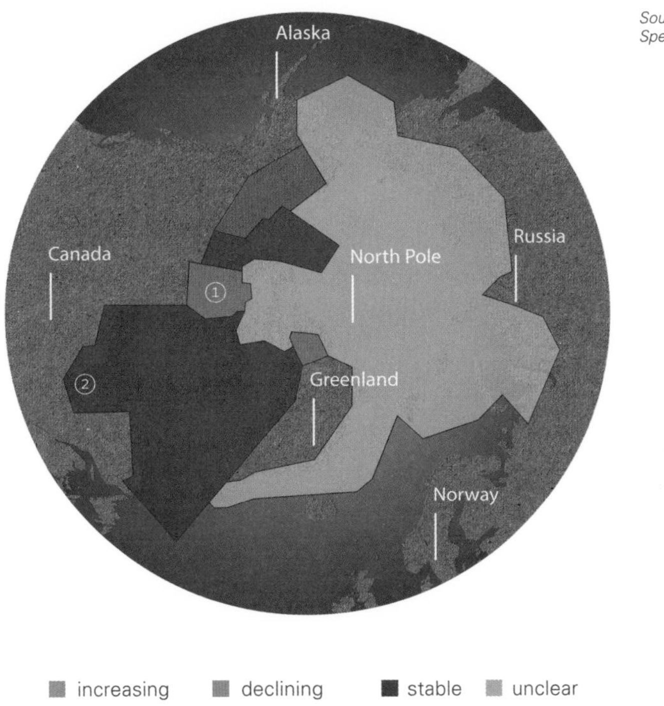

Current status of polar bear populations (simplified). The map shows a snapshot in time rather than any long-term trend. It should be interpreted very cautiously, since the increase in the polar bear population in the north of Canada ❶, for example, is in all likelihood the result of hunting restrictions.[9] Also, the physical condition of the population found in the southernmost habitats in Canada ❷ is deteriorating even though their current status has been assessed as "stable."[10]

Corals

Tropical coral reefs are of crucial importance for humans. The abundance of fish they support serves as a food source, they protect coastlines against erosion by water and wind, and they are an important source of tourism for local economies.[1,2,3] Algae that live on the corals and supply them with nutrients are responsible for their color.[4]

Corals are coming under increasing stress from human-made warming, acidification (p. 68), and pollution of the oceans.[5,6] If the stress level gets too high, the corals expel the algae and their white skeletons become visible (coral bleaching).[4] This can lead to the corals dying because they are no longer sufficiently supplied with nutrients. In 2016, 93 percent of the reefs in the Great Barrier Reef in Australia were affected at times by coral bleaching, and in the shallow-water regions of the Pacific over half of the corals died between February and October 2016.[7]

Richard Vevers, The Ocean Agency

CHAPTER 7

HUMANS

Climate change is already affecting, directly or indirectly, the lives of all 7.8 billion people on Earth.[1] However, the consequences of global warming are not the same everywhere—it impacts people and communities in different ways in different regions of the world. But one thing is clear: As warming progresses, climate change's negative consequences will predominate.[2]

Climate Change and Health

The consequences of climate change affect human health in a variety of ways. Heat stress,[1] for example, can exacerbate heart conditions and circulatory and respiratory diseases, leading to increased mortality rates.[2,3,4] High temperatures also boost the formation of ground-level ozone,[5] which can have a negative impact on health, for example, by compromising lung function.[6,7] Increasingly frequent extreme weather events such as floods or storms pose numerous risks to people's health—potentially fatal injuries are just one example.[8] Heavy rainfall and flooding can cause microbial contamination of rivers and coastal waters, leading to an increased risk of outbreaks of waterborne infectious diseases.[9] Another consequence of climate change is, in the US and some European countries,[10] a prolonged pollen season, which exacerbates the symptoms of respiratory diseases such as asthma and hay fever.[9] Climatic conditions could also encourage the spread of invasive allergenic plants, such as ragweed.[8,11]

Note on the illustration: These figures should be interpreted with caution since it is extremely difficult to calculate the additional deaths that might result from climate change, and no direct causal connection can be established between individual deaths and climate change.

225,000

from malnutrition

85,000

from diarrhea

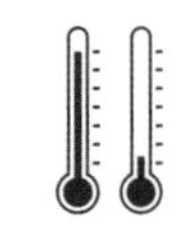

35,000

from heat and cold

30,000

from meningitis

20,000

from vector-borne diseases

2,750

from floods and landslides

2,500

from storms

Additional climate-related deaths worldwide in 2010

Source: Fundación DARA Internacional. Climate Vulnerability Monitor 2nd Edition. A Guide to the Cold Calculus of a Hot Planet (2012)

Other Health Consequences

The consequences of climate change can have a negative impact on mental health. For instance, experiencing extreme weather events can trigger post-traumatic stress disorders,[1] and the worry and uncertainty around climate change can even lead to anxiety and depression.[2] The actual impact of climate change on an individual's health will always depend on their personal circumstances, and how much they are directly or indirectly affected by climate change events.[3] As global warming progresses and more and more people are affected, the risk of adverse impacts on mental health may increase.[4]

What follows are some other ways in which climate change can affect human health, although their manifestation will significantly depend on the region and the level of economic development. Besides the agricultural runoff of nutrients into waterways, higher temperatures can also lead to algal blooms occurring more frequently, for longer periods, and over wider areas[5]—in other words, to a mass proliferation of algae. Some of these species, such as cyanobacteria ("blue algae"), can produce toxins that can enter the human body either via the food chain or by the accidental swallowing of lake or sea water, leading to various illnesses or even death.[6]

Higher temperatures allow bacterial pathogens in food to multiply more quickly. For example, it is thought that higher temperatures could be related to a higher frequency of salmonella poisoning.[7]

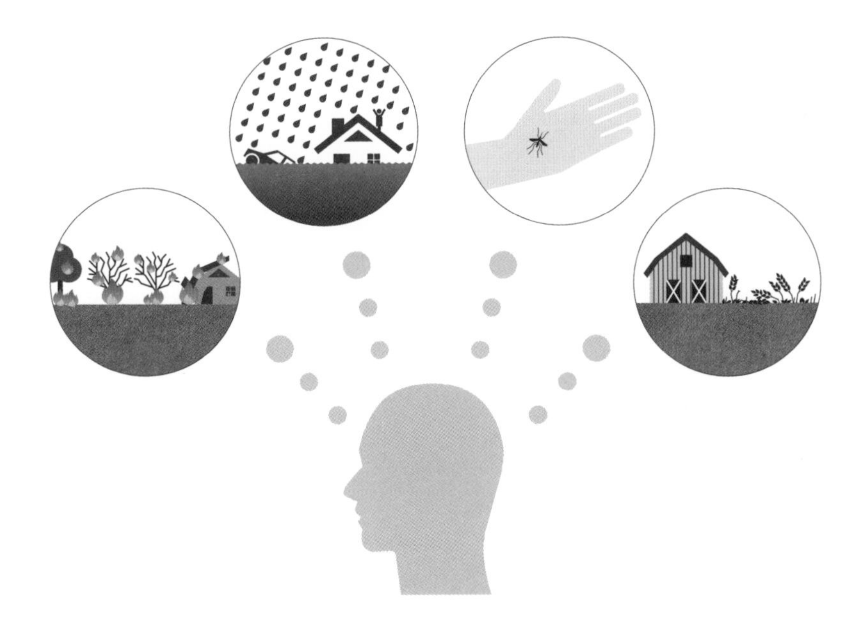

Warmer seawater is likely to increase the concentration of harmful bacteria in the water, a process that may have already contributed to the growing number of infections among bathers in the North Sea as well as the US's Atlantic coast.[8–11] Given the wide variety of effects it can have on human health, climate change is considered the greatest global health risk of the twenty-first century.[12]

Vector-Borne Diseases

Organisms are referred to as "vectors" if they can transmit pathogens from an infected animal or person to other animals or people. Common examples include ticks and mosquitoes.[1]

Climate change is altering the way pathogens spread by vectors.[2,3] For example, as a result of globalization and newly favorable climate conditions, the Asian tiger mosquito (*Aedes albopictus*) has managed to spread into parts of southern Europe over the last few decades.[4,5] Climate change will also make European regions farther to the north suitable for these mosquitoes.[5,6] The tiger mosquito can transmit pathogens such as the dengue and chikungunya viruses.[5]
Higher temperatures must prevail for a certain period for an infected mosquito to transmit the virus. Such conditions facilitate the spread of viruses among the mosquitoes and their transmission to humans when the latter are bitten.[7] Rising temperatures encourage the geographical spread of tiger mosquitoes and will shorten the replication time of the viruses they carry. In combination with globalization, which increases the possibility of tiger mosquitoes being introduced into new countries through imported goods or infected persons (for example, returning vacationers), the risk of disease transmission rises.[2,8]

Source for map: According to the European Centre for Disease Prevention and Control; European Food Safety Authority. Aedes albopictus*—current known distribution: April 2017.*

Spread of the tiger mosquito in Europe
2000
2017
firmly established
sighted, but no overwintering population yet

Cities

Even without climate change, cities have higher air and surface temperatures than their rural or less developed surroundings. Their densely built-up environments and sealed surfaces cause cities to absorb a great deal of solar energy during the day, storing it in their buildings.[1,2,3] Waste heat produced by central heating or air-conditioning systems accounts for additional warming, while gas and particle emissions also trap heat ("haze dome"). The density of buildings, meanwhile, reduces air exchange with surrounding areas, and consequently, when cities get hot, they cool off more slowly.[4] In addition, because there are generally fewer large green spaces, there is only a minimal cooling effect from shade

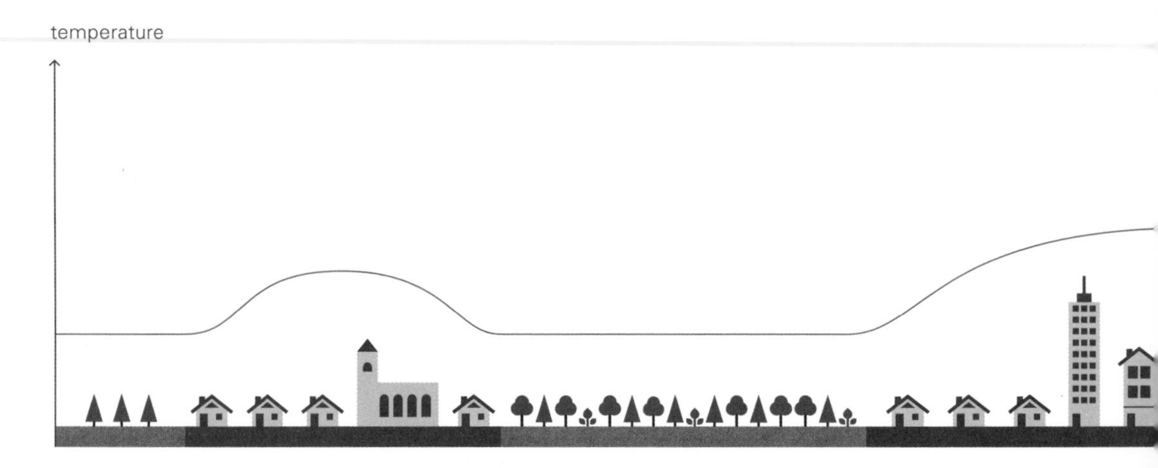

and evaporation.[2,5] As a result, cities can experience temperatures under low-exchange weather conditions that are up to 10°C (18°F) higher than in the undeveloped countryside.[6] This warming is known as the "urban heat island effect," and it occurs particularly at night,[1,6] when most of the energy stored in the structures is released again.[1] Warmer nights can affect how deeply people sleep, thereby diminishing its restorative effect.[7,8] Higher temperatures also lead to greater energy consumption, for example, from air-conditioning units.[9]

Due to the urban heat island effect, the negative health impacts of higher temperatures (p. 104) resulting from climate change look set to become particularly noticeable in cities.[2,6]

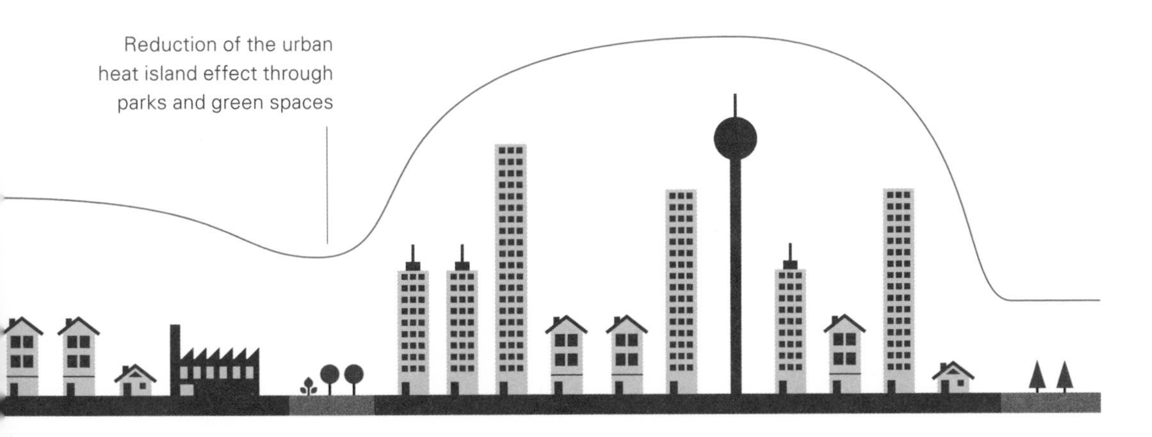

Agriculture

Higher temperatures, greater CO_2 concentrations in the atmosphere, altered precipitation patterns, and other associated weather parameters all influence plant growth.[1] A rise in temperature, up to the "optimum growth temperature," can produce an increased yield for specific crops. However, if this optimum level is exceeded, the crop yield will decline. For example, even a single day of temperatures above 30°C (86°F) can affect the growth of corn and soybeans.[2] Likewise, weather extremes, in particular droughts, heat, and heavy rainfall,[3] have a negative impact on crop yields. Between 2000 and 2007, roughly 6.2 percent of all cereal grains harvested worldwide was lost to drought and heat.[4]

If there is a higher concentration of CO_2 in the air, many plants respond in the short term by reducing how much water they release through their leaves, and simultaneously enhancing their photosynthetic activity. This can strengthen plant growth, provided there is a sufficient supply of water and nutrients—this is known as the "CO_2 fertilization effect."[5,6] The extent to which this effect might

compensate for the reduction in crop yields due to changes in temperature and rainfall is subject to debate.[7,8] In addition, higher CO_2 concentrations in the air in combination with stimulated growth can lead to lower concentrations of nutrients in plants.[9,10]

Climate change may have a positive effect in the world's more northerly regions, such as northern Europe, as crop yields may increase with the rise in average temperatures, due to longer cultivation periods and fewer frosts.[11,12] In the tropics and subtropics, on the other hand, climate change is likely to have a negative impact on crop yields.[13]

In general, it can be said that a rise in the global average temperature in the range of 1–2°C (1.8–3.6°F) (compared to pre-industrial levels) should have a low to moderate impact on crop yields, although there will still be differences depending on the region and type of crop. However, any increase in temperature beyond this will probably lead to a sharp reduction in crop yields.[7,14,15]

Climate Migration

Between 2008 and 2016, weather events—especially storms and floods—drove an average of 21.7 million people to migrate within their own country each year. In 2016, the figure was three times higher than the number of refugees from war and violence.[1] Weather disasters also result in cross-border migration.[2] The majority of those affected are poorer people, generally living in underdeveloped regions where the impact of climate change is particularly harsh. People living in such circumstances usually don't have financial means to adapt to climate change, and state measures often do not go far enough. For these reasons, many people seek to escape its consequences by migrating. Such groups have generally contributed very little to climate change, yet they

Causes of migration

ecological
- weather extremes
- ecosystem services

political
- discrimination
- persecution

social
- education
- family

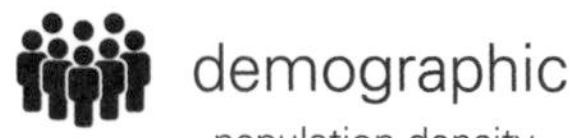

demographic
- population density and population structure

economic
- work opportunities
- salaries

Climate change can have a direct or indirect influence on the reasons for migration.

the are most severely affected by its consequences.[3,4] This becomes especially problematic when people do not even have the option of migrating due to a lack of funds or other reasons.[5]

There are usually several reasons for migration, which makes it hard to ascribe individual migration movements to climate change.[3,6] While it is difficult to attribute a single storm to climate change,[7] for example, climate change may increase the frequency and intensity of such storms, thereby increasing the number of people affected. Consequently, in the absence of adaptation measures, more people will be forced to migrate. Questions about where climate change victims are supposed to migrate to, or whether they should be given special protection, so far remain unanswered.[8]

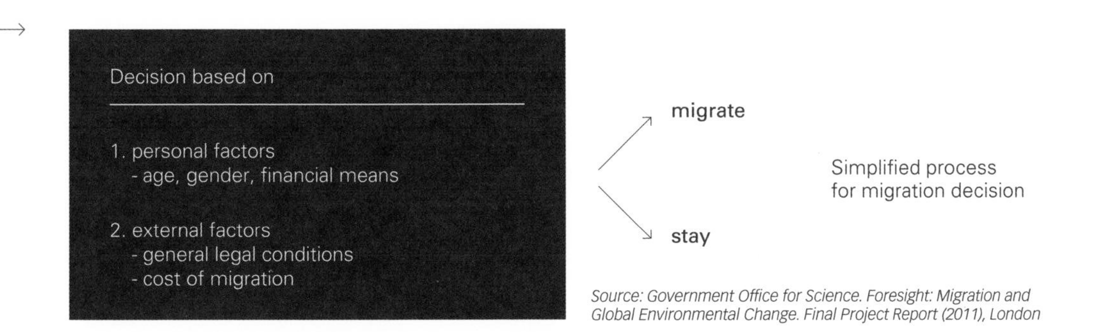

Simplified process for migration decision

Source: Government Office for Science. Foresight: Migration and Global Environmental Change. Final Project Report (2011), London

Tourism

The tourism sector—and especially air travel—was responsible for roughly 8 percent of global greenhouse gas emissions between 2009 and 2013. It is both a driver and a victim of climate change.[1] Climate change could make the Mediterranean regions too hot for many tourists in summer[2]—though, conversely, travel conditions in the regions could become more comfortable in spring and autumn.[3]

Thanks to warmer temperatures, regions closer to the poles and at higher altitudes could extend their travel seasons and prove popular among people who want to escape from the hot weather in summer.[2,4]

Climate change represents an especially grave problem for winter sport tourism, since the rising temperatures mean there are fewer and fewer areas that have guaranteed

More frequent Arctic sea cruises

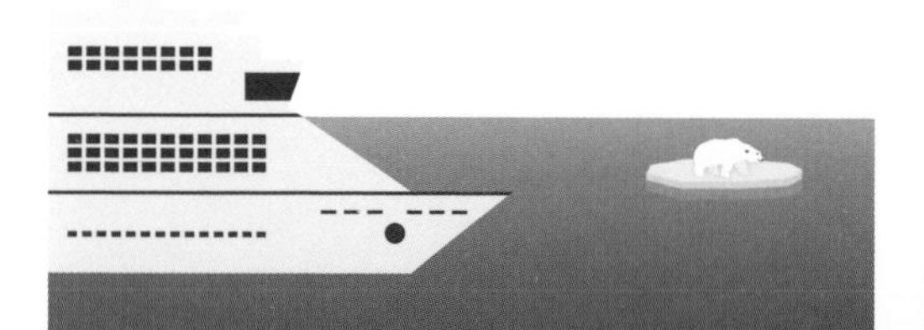

Fewer snow-sure ski resorts

snow.[5,6,7] Many holiday resorts lack their wintry appeal when snowfall is unreliable or patchy.[8,9,10]

Climate change will not hinder global tourism overall, but it will alter the current flow of tourists.[11] The industry will adapt, and worldwide growth in population and prosperity will lead to rising tourist numbers.[12,13] Depending on the type of holiday and the region, the tourism industry will therefore experience both positive and negative impacts should warming be only moderate.[12]

But if global warming becomes more extreme, the losses for the tourism industry as a whole are likely to outweigh any gains,[14] since adaptation measures, such as coastal protection[15] or the production of artificial snow, become uneconomic or ineffective.[5]

Damage to nature

Sinking islands

Economic Costs

Various calculations for annual losses from climate change as a proportion of global domestic product in %

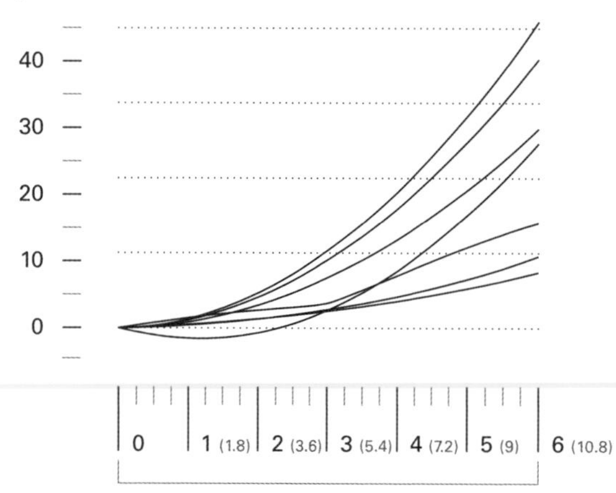

Increase in average global temperature in °C (°F) compared to preindustrial times

Source: © Howard & Sterner (2017) link.springer.com/article/10.1007/ s10640-017-0166-z. CC BY 4.0: creativecommonsorg/licenses/by/4.0.

Climate change generates three different kinds of economic cost. The first is the cost of damage, for example, to real estate or infrastructure as the result of extreme weather events. Second is the cost of adaptation: measures such as the building of dykes or retention basins for flood protection.[1] Third is so-called mitigation costs: measures intended to limit future global warming, for example, switching from fossil fuels to renewables.[2,3]

Putting a precise value on the economic cost of climate change is extremely difficult, as not all costs can be recorded clearly or comprehensively.[4] Calculations will depend on certain assumptions,[5] and some of the costs—such as those that arise from thawing permafrost soil—are very difficult to forecast.[6] The figures in the illustration below should therefore be interpreted very cautiously. Because of the high level of uncertainty, it is also extremely difficult from a purely economic view to determine which protection measures are appropriate, and on what scale. Limiting global warming to a maximum of 1.5°C (2.7°F) will entail extremely high investment costs. Limiting it to a maximum of 3.5°C (6.3°F) would require much lower investment costs,[7] but result in higher costs for the repair of damages.[5] Overall the costs of mitigating global warming are probably much lower than the damage costs that would result from unchecked warming.[2,8] However, it must be remembered that the risk of irreversible damages increases as global temperatures rise.[9]

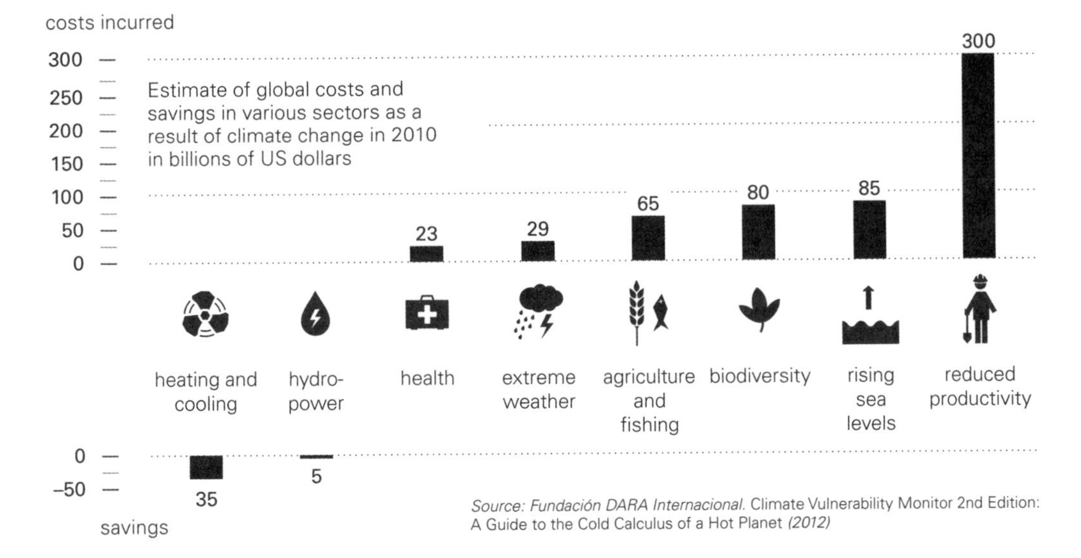

Source: Fundación DARA Internacional. Climate Vulnerability Monitor 2nd Edition: A Guide to the Cold Calculus of a Hot Planet *(2012)*

Conclusion

Climate change is not some far-off reality, and its impact is much more than "just" melting ice and rising sea levels. It's clear from the scientific findings outlined over the last hundred pages that many people and their livelihoods are already threatened by climate change today. It has also become clear that human greenhouse gas emissions are chiefly responsible for the increase in temperature since the start of industrialization. Ironically, this is good news: It's a reminder that we can influence how our climate develops in the future. We are not powerless against climate change.

Climate model simulations show that we can limit global warming if we reduce greenhouse gas emissions ❶.[1] However, since the first UN Climate Change Conference in Berlin in 1995, global greenhouse gas emissions have increased by 50 percent and remain locked at record levels today.[2] If we continue to emit large quantities of greenhouse gases, there could be a further warming of up to 5°C (9°F) by the end of the century ❷.[1] We therefore urgently need to recognize our responsibility. The future of the climate is in our hands, and it is down to us to limit global warming and its consequences.

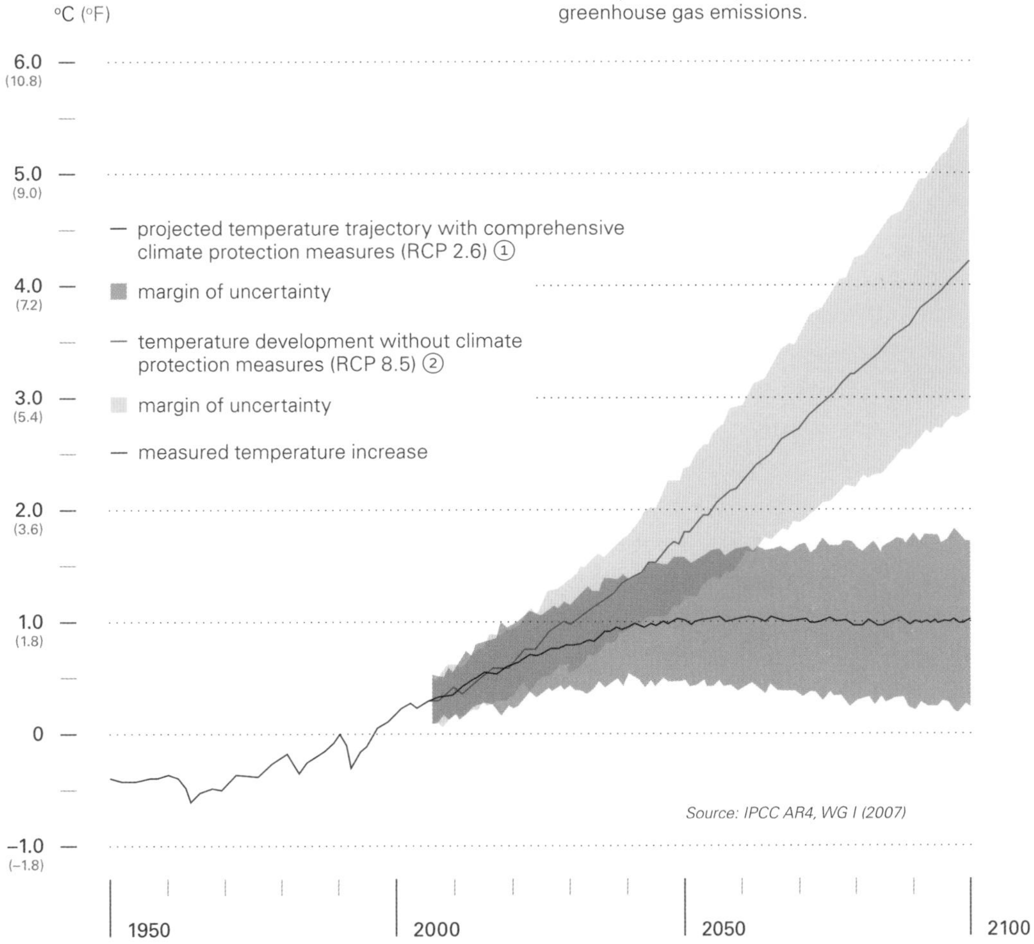

change in temperature
°C (°F)
Projected global average temperature rise up to the year 2100 depending on the volume of greenhouse gas emissions.
6.0 (10.8)
5.0 (9.0)
4.0 (7.2)
3.0 (5.4)
2.0 (3.6)
1.0 (1.8)
0
−1.0 (−1.8)
projected temperature trajectory with comprehensive climate protection measures (RCP 2.6) ①
margin of uncertainty
temperature development without climate protection measures (RCP 8.5) ②
margin of uncertainty
measured temperature increase
Source: IPCC AR4, WG I (2007)
1950
2000
2050
2100

What Next?

Global warming should be contained as far as possible, since the effects of the rising average global temperature are overwhelmingly harmful to us and to our environment. If we are to succeed, it is crucial that we question the true source of greenhouse gas emissions. We must recognize that it essentially comes down to unsustainable human behavior. Human-made greenhouse gas emissions always arise from decisions we make: It is not the car that is responsible for emissions, but rather the person who chooses to drive it instead of taking public transportation or cycling. The efforts of each individual are every bit as important as political action for sustainability on the national and international stage. We are each responsible for rethinking our behavior, and making day-to-day decisions for a sustainable lifestyle and society. We must all become involved in the public conversation, and support sustainability, environmental protection, and climate justice, both at work and in our everyday lives. You will almost certainly meet with resistance, but more often

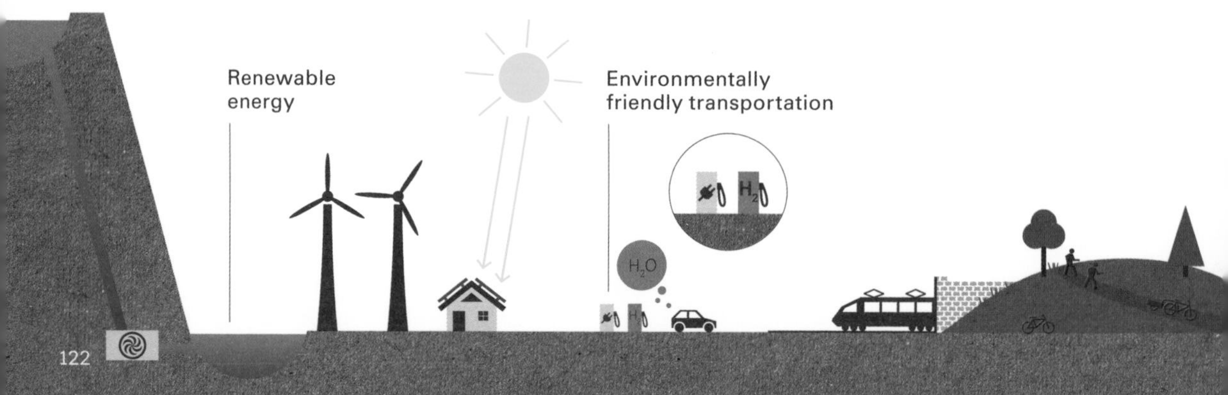

than not you will find assent, and you will meet other people—like yourself—who are motivated to make a change. One thing is clear: "Every person for him or herself" will not save the world. But if we each motivate those around us to protect the natural environment and safeguard our climate—if we all, at every level of society, do what we can—together we will make an essential difference.

—David and Christian

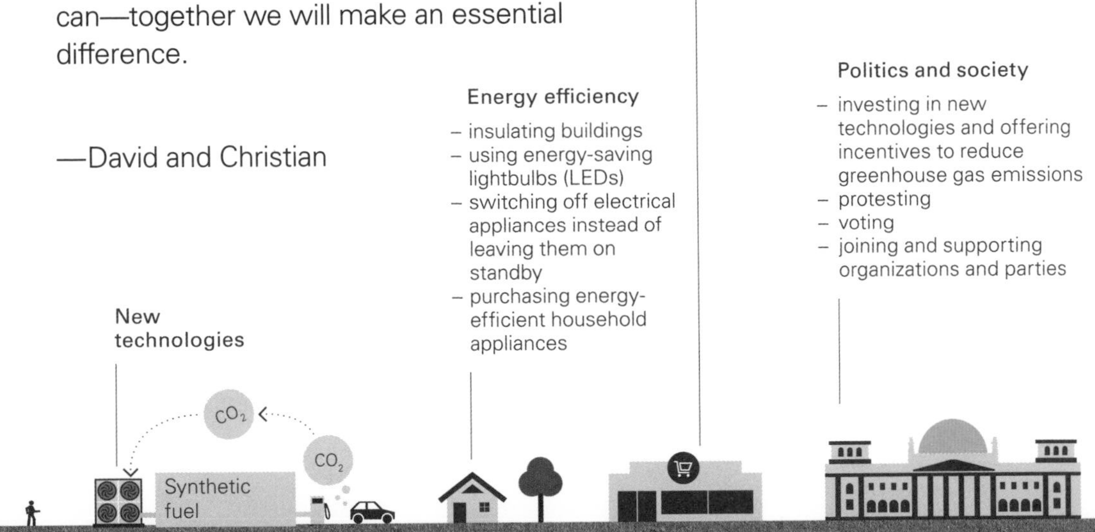

Thank You!

We would like to thank all the experts who have supported us and contributed to the creation of this book. We have benefited from countless insightful dicussions, and many people have been generous with their comments and suggestions. We extend our thanks to:

Prof. Dr. Bruno Abegg | Prof. Dr. Kenneth B. Armitage | Dr. Todd Atwood | Prof. Dr. Herrmann Bange | Dr. Christian Barthlott | Dr. Andreas Bauder | Prof. Dr. Jürgen Baumüller | Prof. Dr. Carl Beierkuhnlein | Prof. Dr. Gerhard Berz | Dr. Tobias Binder | Dr. Boris K. Biskaborn | Prof. Dr. Daniel T. Blumstein | Prof. Dr. Reinhard Böcker | Dr. Benjamin Leon Bodirsky | Frank Böttcher | Prof. Dr. Peter Brandt | Dr. Susanne Breitner | Julia Brugger | Prof. Dr. Nina Buchmann | Dr. Michael Buchwitz | Dr. Paul CaraDonna | Prof. Dr. Martin Dameris | Dr. Annika Drews | Markus Dyck | Prof. Dr. Olaf Eisen | Dr. Georg Feulner | Prof. Dr. Andreas H. Fink | Dr. Mark Fleischhauer | Dr. Achim Friker | Prof. Dr. Martin Funk | Dr. Pia Gottschalk | Prof. Dr. Henny Annette Grewe | Prof. Dr. Christian Haas | Prof. Dr. Wilfried Hagg | Dr. Judith Hauck | Majana Heidenreich | Prof. Dr. Martin Heimann | Dr. Peter Hoffmann | Prof. Dr. Corinna Hoose | Dr. Mario Hoppema | Prof. Dr. Hans-Wolfgang Hubberten | Dr. Amy Iller | Prof. Dr. Kai Jensen | Prof. Dr. Anke Jentsch | Prof. Dr. Konrad Kandler | Dr. Johannes Karstensen | Dr. Stefan Kinne |

Prof. Dr. Gernot Klepper | Dr. Stefan Klotz | Prof. Dr. Peter Knippertz | Dr. Annette Kock | Dr. Peter Köhler | Dr. Martina Krämer | Prof. Dr. Lenelis Kruse-Graumann | Prof. Dr. Michael Kunz | Prof. Dr. Wilhelm Kuttler | Dr. Thomas Laepple | Dr. Peter Landschützer | Prof. Dr. Hugues Lantuit | Dr. Josefine Lenz | Prof. Dr. Ingeborg Levin | Dr. Christian Lininger | Prof. Dr. Karin Lochte Prof. Dr. Gerrit Lohmann | Prof. Dr. Hermann Lotze-Campen | Dr. Remigus Manderscheid | Prof. Dr. Ben Marzeion | Prof. Dr. Katja Matthes | Prof. Dr. Egbert Matzner | Prof. Dr. Marius Mayer | Dr. Hanno Meyer | Prof. Dr. Peter Molnar |Dr. Anne Morgenstern | Prof. Dr. Dr. h. c. Volker Mosbrugger | Dr. Ulrike Niemeier | Dr. Hans Oerter | Prof. Dr. Dirk Olbers | Dr. Marilena Oltmanns | Dr. Daniel Osberghaus | Prof. Dr. Arpat Ozgul | Prof. Dr. Anthony Patt | Dr. André Paul | Prof. Dr. Roland Psenner | Prof. Dr. Johannes Quaas | Dr. Volker Rachold | Prof. Dr. Stefan Rahmstorf | Dr. Maximilian Reuter | Prof. Dr. Mathias Rotach | Dr. Heli Routti | Dr. Ingo Sasgen | Bernhard Schauberger | Lukas Schefczyk | Prof. Dr. Jürgen Scheffran | Dr. Hauke Schmidt | Prof. Dr. Imke Schmitt | Prof. Dr. Jürgen Schmude | Dr. Alexandra Schneider | Prof. Dr. Christian-Dietrich Schönwiese | Prof. Dr. Josef Settele | Prof. Dr. Ruben Sommaruga | Prof. Dr. Christian Sonne | Dr. Sebastian Sonntag | Dr. Robert Steiger | Dr. Christian Stepanek | Dr. Sebastian Strunz | Kira Vinke | Prof. Dr. Martin Visbeck | Dr. Peter von der Gathen | Dr. Mathis Wackernagel | Dr. Frank Wagner | Prof. Dr. Heinz Wanner | Prof. Dr. Hans-Joachim Weigel | Dr. Rolf Weller | Dr. Martin Werner | Prof. Dr. Georg Wohlfahrt | Prof. Dr. Harald Zeiss | Dr. Yves Plancherel

Scientists

Bibliography

You can look up every source cited in this book in the digital bibliography. You will also find interesting suggestions for further reading and websites that cover individual topics in more depth.

The bibliography can be accessed using the QR code as follows:

1. Download a QR code scanner app to your smartphone or tablet.

2. Scan the QR code below.

3. The digital bibliography will open up. Click on the relevant page number and all references cited there will be displayed.

The digital bibliography can also be reached via the following link:

bit.ly/3t4Ukq7

WHO'S BEHIND THE BOOK?

About the Authors

David Nelles and Christian Serrer are economics students at the University of Friedrichshafen. Having noticed that the facts tend to get lost in the fraught public debate about climate change, they decided to get to the bottom of things, once and for all. But, frankly, neither had any desire to trawl through thick textbooks. What they were hoping to find was a book that explained the nuts and bolts of climate change and presented the scientific evidence in a way that was concise and enjoyable to read. After a long and fruitless search, they eventually gave up and instead decided to write it themselves.

David and Christian founded their own publishing house to produce and distribute the original (German) edition of this book in the most environmentally responsible way they could—with eco-minded graphic designers and a sustainable printing press. This North American edition is likewise printed with environmentally friendly vegetable-based ink on responsibly sourced wood fiber. David and Christian hope that this comprehensive guide to the causes and consequences of climate change—as well as the very paper it's printed on—will inspire readers to prioritize environmental and climate protection.